A Propensity to Protect

Butter, Margarine and the Rise of Urban Culture in Canada

W. H. Heick

Wilfrid Laurier University Press

Wilfrid Laurier University Press acknowledges the support of the Canada Council for the Arts for our publishing program. We acknowledge the financial support of the Government of Canada through the Canada Book Fund for our publishing activities.

Library and Archives Canada Cataloguing in Publication

Heick, W.H., 1930–

    A propensity to protect : butter, margarine and the rise of urban culture in Canada

Includes bibliographical references and index.

ISBN 978-1-55458-485-7 (paper)
ISBN 978-0-88920-781-3 (PDF)

1. Margarine industry – Canada – History. 2. Margarine industry – Government policy – Canada. 3. Butter trade – Canada – History. 4. Butter trade – Government policy – Canada. 5. Consumers – Canada – Attitudes. 6. Urbanization – Canada – History. I. Title.

HD9330.M373C3 1991   338.4'766432'0971   C90-095512-0

© 1991 Wilfrid Laurier University Press
    Waterloo, Ontario   N2L 3C5
    www.wlupress.wlu.ca

This printing 2013
Printed in Canada
Cover design by Rick McLaughlin

A Propensity to Protect: Butter, Margarine and the Rise of Urban Culture in Canada has been produced from a camera-ready manuscript supplied by the author.

Every reasonable effort has been made to acquire permission for copyright material used in this text, and to acknowledge all such indebtedness accurately. Any errors and omissions called to the publisher's attention will be corrected in future printings.

No part of this publication may be reproduced, stored in a retrieval system or transmitted, in any form or by any means, without the prior consent of the publisher or a licence from The Canadian Copyright Licensing Agency (Access Copyright).

For an Access Copyright licence, visit www.accesscopyright.ca or call toll free to 1-800-893-5777.

CONTENTS

TABLES AND GRAPHS v

ACKNOWLEDGMENTS vii

INTRODUCTION 1

PART ONE
The Kingdom of the Cow:
The Life of the Ban

CHAPTER ONE
Imposition of the Ban on Margarine: 1880-86 6

CHAPTER TWO
Exclusion: 1886-1914 22

CHAPTER THREE
Momentary Easing: 1914-19 30

CHAPTER FOUR
Renewed Exclusion: 1919-24 43

CHAPTER FIVE
Holding the Line: 1924-45 53

PART TWO
The End of the Ban

CHAPTER SIX
The Fight to End the Ban: 1946-48 62

CHAPTER SEVEN
Reference to the Courts: 1949-50 79

PART THREE
Margarine Prevails: 1950-87

CHAPTER EIGHT
The Legislative Maze 94

CHAPTER NINE
Government Involvement since 1950 105

CHAPTER TEN
Economics of the Butter and Margarine Industries 117

CHAPTER ELEVEN
**Margarine, Butter, Nutrition and Colour —
"Holy Cow, It's Still a Holy Cow!"** 134

CHAPTER TWELVE
Marketplace Competition 147

CONCLUSION 161

APPENDICES 165

NOTES 175

BIBLIOGRAPHY 203

INDEX 221

TABLES AND GRAPHS

Table		
2-1	Population Increase Re Urbanization	26
3-1	Recommended Daily Food Needs (1917)	34
10-1	Purchasing Power (Wholesale) of Agricultural Products	118
10-2	Indexes, Creamery Butter Production and Farm and Retail Prices	120
10-3	Profitability of Various Uses of Raw Milk	120
10-4	Farms Reporting Milk Cows, 1961-71	121
10-5	Per-Capita Weekly Food Purchases by Family Income Quintile Group, Canada, 1969 and 1974	125
10-6	Major Margarine Developments, 1890-1967	126
10-7	Land Seeded to Domestic Oil Seeds	128
10-8	Percent Distribution of Oils Used in Margarines in Canada	129
10-9	Comparative Economic Efficiency	129
10-10	Census Farms Reporting Milk Cows	131
10-11	Farm Data by Total Capital Value — Canada	132
10-12	Quarterly Wholesale Butter Prices in Canada and in Selected Countries, 1958	133
11-1	Nutritive Value, Edible Portion of 100g (1980)	136
11-2	Actual Contribution to Fat by Major Foods in Canada, 1960-75	139
11-3	Percentage Contribution to Total Fat by Major Foods in Canada, 1960-75	140
11-4	Per-Capita Annual Supplies of Food Moving into Consumption	141
11-5	Percent Distribution of Margarine Samples Containing Less than 5% to More than 20% Cis, Cis, Methylene-interrupted Polyunsaturated Fatty Acids (CCMl)	142
12-1	Average Retail Price Per Pound	150

12-2	Butter and Margarine Use in Ontario, 1957 and 1965	154

Graph
6-1	Butter Retail Price Index	69
10-1	Average Retail Prices and Price Relatives for Selected Time Periods, Canada	123
10-2	Margarine/Butter Price Comparison, 1950-85	124
11-1	Indexed Consumption Per Capita of Selected Food Commodities, Canada, 1911-73	135

ACKNOWLEDGMENTS

The history of margarine in Canada opens a new topic in Canadian history. The thesis is that the gradual acceptance of this new food over the first century after Confederation was generated by the historical factor of the shift from a rural society to one with predominantly urban perspectives.

Although topically new, the writing of this history has used traditional sources: federal and provincial government documents, the papers of individuals and organizations, newspapers and magazines, and personal interviews. The generous support of the archivists, federal and provincial, across the country, is gratefully acknowledged. The officers and staff of the Institute of Edible Oil Foods and the Dairy Farmers of Canada were also most willing to help in any way asked. Two prime areas of limitation developed. It has not been possible to circumvent confidentiality relating to certain government documents. Secondly, queries to manufacturers of margarine, such as Schneiders, met with the response that records of a historical nature were not maintained in such a way as to be of help with this book.

Colleagues and friends have read parts or the whole of the manuscript: Jacques Goutor of King's College and Suzanne Zeller of Wilfrid Laurier University have performed immeasurable service by wielding sharp and wise editorial pencils. Others have made sound criticisms of the whole manuscript: Nicole Begin-Heick of the University of Ottawa, Hans Heick of the Ottawa Children's Hospital, Heather Heick, Lisa Keeler and Scott Keeler and Andrew Walesch of Conestoga College. Readers of segments of the book have been Jane Colwell of Wilfrid Laurier University, Arthur Gaull, John Gaskell, Reginald Haney of Wilfrid Laurier University, Margaret Hyndman and Farley Mowat. Terry Wuester of Victoria also helped with legal research. Bryan Stortz of Kitchener served exceedingly well as research assistant for a summer.

Jean Gourlay performed magnificently in deciphering my handwritings with a critical eye and in offering friendly encouragement. Olive Koyama's, Doreen Armbruster's and Elsie Grogan's sharp eyes caught numerous problems.

This book has been published with the help of a grant from the Social Science Federation of Canada, using funds provided by the Social Sciences and

Humanities Research Council of Canada. Wilfrid Laurier University also contributed generously to the costs of research and publication.

To acknowledge, with sincerest appreciation, the support of my wife and friend, Marg Heick, is the happiest task of all.

— *Courtesy Provincial Archives of Alberta, Pollard Collection, P4127, 78.63/1331*

INTRODUCTION

To say that food is one of the necessities of life is to belabour the obvious. Historians have, for the most part, left the study of what people eat to the nutritionists, biologists and chemists concerned with the physiological aspects of food. Only in recent years have historians become interested in the subject, and established the close relationship between the patterns of dietary habits and the course of economic and social change.

Beside the importance of studying food as a socio-economic phenomenon, there are two more reasons for the present work. To date there has been no full-scale study of margarine's role in Canadian history. Even those monographs that deal with the international scene either pay no attention to Canadian events[1] or err when they do.[2] Secondly, the history of margarine in Canada is somewhat unusual since this country was one of the very few national states to ban the product outright, rather than to control or restrict its use through taxes or regulations.

Margarine originated in France and was developed in response to the need to offset the large deficiency in the supply of fats for the diet of the poor, in particular urban, working-class people. Western Europe was too densely populated to support dairy and cattle herds large enough to meet all its fat needs. The Industrial Revolution, with its development of railways and an extension of industry, also generated a vast new field in which fats and oils were required. During the Franco-Prussian War, Napoleon III offered a prize for the invention of a satisfactory substitute for butter, stimulating thereby the research that had been underway since early in the century. The prize was won in 1879 by Hippolyte Mége-Mouriez. His process was essentially that of putting fresh beef suet through a factory process similar to that by which a cow produces milk. The product was then churned with milk in order to impart some of the flavour and consistency of butter.

Mége-Mouriez's teacher, Michel Chevreul, in 1813, had coined the word "margarine" for a fat-like substance he had prepared. "Margarine" was derived from the Greek word for "pearl," since the crystalline substance had a pearly lustre. Chemically speaking both oleomargarine and margarine are misnomers, as it has since been established that Chevreul's substance was really a mixture of two acids — stearic and palmitic.[3] In the Canadian political, economic, and social realms the two terms have been used indiscriminately. To the degree that sense can be made of popular usage, oleomargarine referred to the product when it was derived primarily from animal fat, and margarine applied when vegetable oils were the base ingredients.[4] In this study, the term margarine has been used throughout, except when some technical differences between the two products were being developed; oleomargarine has been maintained in quotations.

The history of margarine is a prime example of applied technology meeting the changing needs and desires of the consumer. Margarine was invented as a cheap substitute for butter so that the poorer elements of society could have some spread on their bread. A key element of margarine's history is its continuing competition with butter for the patronage of various consumers.

The agrarian myth stipulates that life in a rural setting creates the soundest environment in which a person can develop. Thus butter came to symbolize the virtues of the rural lifestyle, while margarine was perceived to exemplify a less attractive urban way of life. The producers of butter argued that their product was the best because it came from a naturally produced ingredient with almost no additives. The myth of milk as the perfect food carried over into implications of the natural goodness of its by-product — butter.

The butter industry did little to meet the challenge of margarine through product innovation. Instead, it concentrated its attack on the issue of fraud. Building on the public image of margarine as suspect, the butter interests fought to create and maintain Canada's 1886 ban on margarine as a protection against deception of the consumer. The perceived uniqueness of butter only intensified the conflict over such peripheral issues as colour.

Katherine Snodgrass makes three points as to the reaction to margarine in the Western industrialized world. First, European countries in which dairying was strong, such as Denmark and Holland, had foreign outlets; historically, an important one was Britain. Often the Europeans could get a better price for their butter in foreign markets than at home and there was an urgent necessity to earn foreign exchange. The Danes even went so far as to subsidize the use of margarine domestically in order to have more butter for export. Secondly, the rights and interests of consumers were perhaps given more attention in Europe than in newer countries where the population was less dense. In the United States and the British domains overseas the rights of producers were of greater concern to the body politic. Thirdly, the greater portion of the working classes in most European countries was urban and had developed effective means of political action. As a result the state had to follow policies stressing cheap food for the consumer. Thus in Europe margarine was viewed as a valuable, cheaper substitute for a costly food staple. Legislation needed to deal only with the fraud aspects. In North America margarine was seen much more as a threat to the butter producers, resulting in an overreaction to the possibilities of fraud.[5]

Snodgrass's points illustrate why the North American response was more stringent than the European. But they do not explain the extreme Canadian reaction. In nineteenth-century Canada, the natural, universal conservatism of the agriculturist was fortified by economic and technological developments. In the Confederation period, especially in Ontario, farmers were working through a change in emphasis from wheat as the staple commodity, to mixed farming in which dairying played a major role. Caught in the middle of great change, these Canadians reacted by seeing margarine as a threat to the survival of their rural values and lifestyle. Margarine could not appear as anything but a danger, even though the basic arguments in its favour (which were ultimately to win the day, decades later) were already being expressed in the debate of the 1880s.

Mid-nineteenth-century Canadian society, parochial for the most part, basically rural in nature and believing in agrarian values, had rejected margarine outright in order to protect the interests of its economic and social mainstays. By the mid-twentieth century, Canada had become a mixed but predominantly urban society. The middle-class leadership of the generations following World War II was raised for the most part in an urban environment and identified less strongly with the rural life. Better knowledge of nutrition helped develop a more sophisticated appreciation of the role of diet in life. Butter no longer needed to be, nor was it regarded, as one of the key elements of Canadian diet. Margarine was to be accepted as a food in its own right, not solely as a substitute for butter. Its use would protect the urban living standard.

In the spring of 1946, eleven months after Hitler had been defeated in Europe and Canadians could commence to appreciate the rewards of victory, Senator W.D. Euler rose to speak: "I feel pretty strongly as to the need of protecting the interest of the Canadian consumer,"[6] he began. Euler wished to end the prohibition on the manufacture, sale, and importation of margarine, for he felt that Canadians ought to have a free choice between butter and margarine.

Euler's action was a turning point. April 1946, marked the diamond anniversary of Canada's ban on margarine. Since 1886, legislative action had gradually tightened the ban. The one break in this pattern came in 1917, when it was temporarily lifted in order to help meet wartime needs for food. The ban was reimposed in 1923 as the dairy industry, one of the most powerful lobbies in Canada, asserted its will. In the post-World War II years, the desire of an urbanized society for a higher standard of living instigated the ending of the federal ban, and a shift of responsibility to the provinces. Since 1950, provincial response has varied. Most have moved to a position of no restrictions, except for pure food regulations. Only Ontario's and Quebec's controls remain the exception. Margarine, becoming a distinct food in its own right, competes with butter for consumer allegiance in the Canadian marketplace. The contemporary debate over the impact of nutrition on personal health has pulled both foods into the limelight.

This study reflects the dynamic nature of Canadian society and its attitudes, and the transformation from an agrarian outlook to an urban perspective as its dominant interest groups changed. Margarine itself has been a noted example of the possibilities inherent in food technology to meet the changing needs of society. It is also an example of consumer movement to an alternate food when a traditional food was unable to meet consumer demand and price needs. In the public realm the butter/margarine competition illustrates the willingness of Canadians to use the state to achieve their goals.

PART ONE

The Kingdom of the Cow: The Life of the Ban

CHAPTER ONE

Imposition of the Ban on Margarine 1880-86

The British North American colonies in the mid-nineteenth century were strong rural, agriculturally oriented communities. In the Maritimes, the rural proportion stood at 91%; in Lower Canada, 85.1% and in Upper Canada, 86%.[1] While the farming population of the latter two provinces declined between 1850 and 1870, from about 66% of the total occupational spectrum to 55.6%,[2] it actually increased in absolute numbers by some 155,000.[3] In both Quebec and Ontario agriculture was the key element in the economy. The advantage Ontario possessed over its sister province in the St. Lawrence Valley lay in the "vastly superior productive capacity of Ontario agriculture."[4] In the Maritime provinces agriculture and a rural way of life were dominant. Specialization in dairying there did not begin until the 1880s, some two decades later than in the St. Lawrence provinces.[5]

In this pre-industrial age a major journal, *The Agriculturist and Canadian Journal*, held those who tilled the soil to be "'first-class' in the noblest and best sense.... The sturdy yeomen are the true conservatives of society ... the substratum ... of the social fabric.... It has been so in all time, in all other countries; it is so in ours."[6] Yet the need for cooperation between agriculture and industry was also recognized. By the 1840s criticism of the farmer, which was to permeate the history of the butter/margarine rivalry, began to be heard. The *Journal* complained that farmers were suffering from a lack of scientific knowledge, from "our contracted means and ... from a want of inclination."[7] Many Canadian farmers were even then failing to seize the opportunities presented by more scientific agricultural knowledge.

From at least the turn of the nineteenth century, wheat had been the staple product for most Upper Canadian farmers. This remained true as long as land was abundant relative to labour and so long as the original fertility of the soil held up. Once fertility was depleted, and intensive farming was required, wheat production had to give way to the grain coming from the newly settled, more fertile American West. The shift away from grain production intensified when the railroads made transport of animal products to the growing urban markets of Canada and the United States more profitable than the shipping of wheat. The first step in the transformation of agriculture came between the late 1850s and the late 1860s, the second after the late 1880s.[8]

The shift from wheat to mixed specialty agriculture brought about mechanization, diversification, and a re-evaluation of farming methods, including crop rotation. Mechanization could be achieved relatively easily. Implements were not expensive; in fact their price was falling, their cost representing no more than 6% of total value of farm assets until well into the twentieth century. But better livestock and buildings were expensive. It was wheat which provided the capital needed for investment in machinery, livestock and better buildings.[9] Additional factors in stimulating change were the advances made in transportation, especially rail and steamship, which reduced costs for products being shipped to overseas markets. Thus, Canadian goods came to enjoy an improved competitive position, particularly in the all-important British market.

In Ontario agriculture, up to the 1860s, wheat was the "engine of economic growth"[10] not only for the farmer, but also for the urban dweller, as it stimulated commerce and industry. The failure of the Quebec agricultural sector to have in hand a profitable crop such as wheat was the chief reason that province did not achieve Ontario's level of prosperity in the pre-Confederation era. During the 1850s and 1860s, an Ontario farmer had cash sales three to five times those of a Quebec farmer. (See Appendix 1.) As for the Maritime provinces, their story parallels that of Quebec more closely than that of Ontario.

The agricultural advantage held by Ontario at mid-century showed itself more clearly a decade later in the greater production of butter in Ontario, in contrast to Quebec. Of Ontario's eighty-four electoral ridings thirty-five produced more than 700,000 pounds per year, while only twelve of sixty-one ridings did so in Quebec. The dairy revolution affected the Maritimes to a much lesser degree. In Prince Edward Island and Nova Scotia that plateau was reached by one riding in each province and four ridings in New Brunswick. No western riding reached this high level of productivity in butter. (See Appendix 2.) Of the forty-six butter factories in Canada in 1880, half were in Ontario, twenty-two in Quebec and one in British Columbia. The total value of creamery butter produced in 1880 in Ontario was 59% more than the Quebec total.[11] Ontario's butter industry was much more vital than that of Quebec and the Maritimes. As a result, the province's Members of Parliament were to be more sensitive to the appearance of margarine in the marketplace.

In the pioneer era, neither cheese nor butter were profitable commodities for a farmer to produce. There was no real local demand as many villagers kept their own cows and made their own butter and cheese. The usually inferior quality of both, as produced on the frontier, kept selling prices low. Until the 1860s, prices received for butter or cheese showed little change over time. There was more profit in wheat and the farmer stayed with it until the soil became exhausted.

As wheat production gradually declined, beginning in the 1860s, dairying came to be perceived as a godsend, although whether to produce butter or cheese remained a local or even a personal choice. Less susceptible to the vagaries of climate and more stable than wheat in income production, dairying also used a steadier supply of labour and even, at peak times, less labour than wheat. The Quebec Commissioner for Agriculture saw it as, "without doubt, the greatest agricultural progress that has taken place for the last fifty years."[12] Quebec's location on the banks of the St. Lawrence also made it possible to reach the British market two weeks ahead of cheese or butter from western American states.[13]

Near growing urban centres, fluid milk dairying became important around 1880. With dairying the farmer could also address the major problem of soil depletion. Better crop rotation and the availability of great supplies of manure were appreciated as cures for the damage done to the soil by wheat. In 1884, the select committee of the House of Commons looking into the conditions in agriculture concluded that "Dairy farming thus induces a happy concatenation of causes and effects which come very near solving the great problem of the regeneration of agriculture where it is needed."[14]

Producing an acceptable quality of butter was a great problem for the pioneer farmer. Housing for cows was scarce, feeds were home-grown and few in number, and standards of excellence were virtually unknown. Cows were usually dry in winter, a foretaste of the cyclical production problem that would plague both the butter and cheese industries. The task of turning each small lot of milk into butter without either proper technical knowledge or equipment led to inconsistencies in the final product. Catherine Parr Traill argued that instead of being "surprised that there is so little really fine butter sent to market, the wonder should be that under such disadvantages, there is so much."[15] Even after Confederation a lack of uniformity and tendencies toward oiliness, softness, and rancidity prevented Canadian butter from building a strong reputation in international markets.[16]

This need not have been the case: Canadians could be good dairymen and could produce good quality butter. In 1834, the *British American Cultivator* related the case of a Yorkshireman who rented a farm north of Toronto. On the income from ten cows and the sale of their calves he had more than enough to cover a rent of £50. "We examined his stock and found them in comfortable winter quarters, with an abundance of good hay and cut oats, sheaf and bran before them, and a good supply of clean straw under their feet for bedding."[17] Before Confederation, buyers from the American and British markets sought out butter from Lower Canada's Eastern Townships and from eastern Upper Canada. To generate increased production, more information, more care and, most importantly, the establishment of factories was required. The first butter factory was set up in Ontario in 1867.[18] The introduction of the cream separator in the 1880s made possible the extraction of butterfat on the farm and was another technological step forward. The old procedure had required the hauling of the raw milk to the creamery, a task involving the movement of six times the quantity of fluid.[19] The final technical factor—a simple and accurate way of measuring butterfat—was adopted from U.S. sources by Canada in the late 1880s.[20]

At first, butter production was absorbed largely by consumption on the farm and by the urban domestic market. But, by the early 1870s, butter represented nearly 10% of the total value of Canadian agricultural output. However, this level was to increase only to the 11-12% range and to remain there until the turn of the century.[21] In the area of exports, the expansion was also evident. For unexplained reasons, 1858 seems to have been the take-off year when exports stood at 3,721,200 pounds. By 1871, exports had increased five-fold. The peak year came in 1880 (19,887,203 pounds), followed by a rapid decline by 1883 (8,106,417 pounds).[22]

What had happened? Following the experience of 1880, optimism had been high. As the Ontario Agricultural Commission put it, in 1881: "only an intelligent determination to consult their own interests as well as the wants of the market... is needed to make Ontario as fine a butter-producing country as any in the world."[23] Yet within three years butter exports had declined sharply. Although there is some dispute as to the ratio of "good" to "poor" Canadian butter[24] it was agreed that while Canadians did produce some excellent butter, much more was possible and necessary. Eastern Townships and Brockville butter could be sold on reputation alone, but the poorer half of Canadian butter was considered in Britain as useless for anything other than to grease axles or smear sheep. A comparison of prices illustrates the problem. In the 1870s, common Ontario and Quebec butter sold in Britain for 7 pence per pound on average, while U.S. butter sold for 12 pence, French for 14 1/2 pence and Irish for 15 pence.[25] A decade later, the picture was basically the same: whereas the best butter was bringing 120-140 shillings per hundredweight, Canadian butter was bringing 60-120. Quebec farmers alone were losing between five and twelve million dollars annually.[26]

Even though it had been able to secure a solid foothold in the British market, the Canadian butter industry had not been able to overcome the factors of a huge geography, a widespread lack of capital and poor organization to make the production and marketing improvements needed to hold its own, as the industry in competing countries took up the challenge of an almost bottomless British market. An English butter importer offered a compendium of errors: a lack of uniformity in colour, taste, texture, size and shape of package, and standards of cleanliness as a result of not concentrating butter manufacture in factories; a failure to use fine English salt instead of coarse Canadian salt; and a failure to use smooth-sided kegs.[27]

Thus the contention that Canadian farmers shifted from wheat to mixed farming because they preferred greater security of income to the prospect of higher profitability, if the latter involved risk, is borne out.[28] As it developed, Canadian society tended to stress security and survival, not innovation. Conservative by nature and insecure as to profits, Canadian farmers were not prepared to be innovative while making the switch to a modern dairy industry.

Such reluctance existed even when the dairymen and butter producers were being told how their competitors were making better quality butter or using substitutes such as margarine to improve their capacity to export butter. By 1880, Australians were shipping their butter to Britain in hermetically sealed cans, rather than rough wooden kegs.[29] European butter producers, such as the Dutch and the Danes, were energetic in their efforts. The Dutch, for example, doubled their exports to Britain between 1872 and 1880. Both countries were encouraging, by means of government subsidies, the domestic consumption of margarine in order to have more butter to export.[30] Germany, Holland and the United States were shipping margarine to Britain: cheaper by half, it was preferred to the poorer grades of butter, no matter who produced them.[31] In contrast, Canadians opted to use the federal government to protect their domestic market from the competition of butter substitutes. This was so even when there would have been some economic advantage to the agricultural sector from the use of animal (in particular, beef) fat in the production of margarine.

In its first ten years, the federal Department of Agriculture had concentrated almost exclusively on the encouragement and supervision of immigration. Support of agriculture was left to the provincial departments. By order-in-council of April 17, 1877, the Dominion Council of Agriculture was established with representation from all provinces. The Council's task was to investigate what could be done at the national level to bolster the fortunes of the various sectors of agriculture. The first concern of the standing committees, one of which centred on dairying, was to collect data on which to base its reports.[32]

At the provincial level concerns about standards of production were uppermost. For example, in 1883, the Prince Edward Island House of Assembly passed legislation to protect and encourage the manufacture of butter and cheese. The main concerns were standards of purity of product, and of cleanliness of farm and factory facilities. No reference to margarine or any other butter substitute was made in the bill or the debate, although during that year, the first margarine factory in Newfoundland began operations.

The British North America Act contains no specific reference to food and drugs either in section 91, which sets out the areas of legislative competence of the federal parliament, or in section 92, which delineates the fields of jurisdiction for the provincial governments. Section 91, head 2, regulating trade and commerce, and head 29, giving the federal government jurisdiction over subjects expressly excepted in the enumeration of subjects assigned exclusively to the provinces, provide the means by which the federal government could intervene in the areas of food and drugs. Similarly, section 92, head 13 (property and civil rights) and head 16 (matters of a local nature) could be pointed to as the base for provincial action. Section 95 dealing with agriculture, the one other section which might be used to gain some clarification of this question, establishes concurrent responsibility. Therefore, in 1886, it was constitutionally possible to use either level of government to gain one's ends.

However, in 1867, the imperial/colonial/local structure of government had been altered, as far as British North America was concerned, by the addition of a new level of jurisdiction. As a result it was expected that the significant action after 1867 would occur in Ottawa. As well, anyone personally ambitious to be a mover and shaker would view the bigger Ottawa stage as the logical one for his activities. It was not mere historical accident that men such as Cyril Archibald, dealing with the prevention of food adulteration in 1878, and George Taylor, aiming to regulate manufacture and sale of butter substitutes in 1886, were members of parliament and not members of provincial legislative assemblies. Their work brought about a single legislative response at the national level which imposed a new condition on the entire nation, even though only one province, Ontario, was immediately affected.

The first specific evidence of the arrival of margarine into Canadian politics came in 1878. Cyril Archibald, member for Stormont in eastern Ontario's butter belt, introduced a bill to amend the 1874 legislation to prevent adulteration of food, drink and drugs. The earlier legislation, the first concerning food and drug regulation, dealt with "adulteration by the addition of improper substances" and was in response to widespread public concern.[33] Since it made no reference to "adulteration by substitution" it may be assumed that margarine had not yet made any significant inroads in Canada. Because Archibald's bill was concerned with "adulteration by substitution" one can assume that the situation had changed and that he wished to deal with the problem before it grew too great: "This article has not been manufactured in Canada as yet. . . ." His justification for the action was protection of consumers: "some steps should be taken to prevent the imposition of this article upon purchasers as [if it were] genuine butter." All packages of the substitute would, under the law, be required to have placed on them, at both the wholesale and retail level, an "oleomargarine" label. Thus butter consumers would be secure from this cheaper product, and the honest producer of margarine would be protected from the fraudulent manufacturer. Archibald did not argue on health grounds, since he recognized that when properly made, margarine "required almost an expert to detect the difference in appearance . . . between the genuine article [butter] and the substitute." He used the experience of the British and the Americans as evidence of the necessity for this legislation.[34] His private member's bill achieved passage without debate, demonstrating a lack of information and interest among the members.

The media served as reporter, not catalyst, of this political initiative. The respected Ontario agricultural journal, the *Farmer's Advocate*, recognized that margarine "is in itself innocent. It is not butter; never can be." The *Advocate* perceived the wholesalers and retailers as responsible for the adulteration. No one would feel sorry for people who, in their greed, destroyed their reputation and business when their actions were discovered. "The only way for dairymen to fight oleomargarine is the true way. Make a better article than the 'fat butter.' "[35]

The use of margarine developed slowly. It was not until two years later, in 1881, that the next reference appeared in the *Advocate*. It carried a letter which described the unsavory substances the writer had found in margarine and condemned it under two categories that would remain part of the ensuing argument: poor quality of goods and protection of the public from fraud.

> I have examined a large number of specimens of oleomargarine and have found in them organic substances in the form of muscular and connective tissue; various fungi, and living organisms which have resisted the action of boiling acetic acid; also eggs resembling those of the tapeworm. . . . The French patent under which oleomargarine is made requires the use of the stomachs of pigs, or sheep. This is probably the way the eggs get in. I have specimens of lean meat taken from oleomargarine. There can be no question that immense amounts of oleomargarine are sold and used as pure butter. I regard it as a dangerous article, and would, on no account, permit its use in my family.[36]

Four years later the *Advocate* carried a series of articles about margarine. In the spring of 1884, it was reported that the butter/margarine competition had been joined by "butterine," a mixture of butter and deodorized lard. The report advised the Canadian dairy industry to force proper identification of products; to move to winter dairying to even out annual production and thus seasonal prices of butter; and to manufacture butter in factories, to raise the average quality of butter above that of its competitors.[37] Two months later, the *Advocate* shifted its editorial position by calling for the banning of margarine, as the American states of New York, New Jersey and Missouri had recently done.[38]

Non-media pressures against margarine in the marketplace were much slower in developing. In 1882, the Dairymen's Association of Western Ontario, at its annual meeting, listened to a Mr. Gilbert of the Utica, New York *Herald* arguing that margarine had not had the effect of lowering the price of butter, but no action followed.[39] A questionnaire sent by the select committee of the House of Commons in 1884 to obtain information about the agricultural interests of Canada made no reference to the question of butter substitutes.[40] Two years later, and less than one month before the Commons held its first major debate on the topic, a meeting of the dairymen's association in Bayfield, Nova Scotia generated no comment or action concerning margarine.[41] In the forthcoming Commons debate, the Minister of Finance was to state that he had no recollection of a single anti-margarine petition from any agricultural association.[42]

Activity, when it did come, demonstrated the catalytic nature of the protective impulse. At the Ontario Agricultural and Arts Association's annual meeting in March, 1886, the first action was taken: a resolution requesting the federal government to prohibit importation, manufacture and sale of margarine, except when the package was obviously labelled so as to allow the consumer to differentiate the product from butter;[43] in reality the same intent as that of Archibald's 1878 bill.

The option of stringent parliamentary action developed on March 12, 1886, when George Taylor, a merchant and Conservative member for South Leeds, another important Ontario butter and cheese producing area, gave notice of a private bill to regulate the manufacture and sale of margarine and other butter substitutes. His constituents had lobbied him to do something to counter the entry of margarine from the United States.[44] The catalyst was the news that New York entrepreneurs were contemplating the construction of a half-million-dollar factory in Montreal to produce margarine.[45] Having failed to obtain either a satisfactory response or a counter-proposal from the government, Taylor decided to take the private-member bill route to prod the government into action.[46]

The government of Sir John A. Macdonald was therefore responding to various pressures when, on April 2, the Minister of Inland Revenue, John Costigan, placed on the order paper notice of legislation dealing with the matter of margarine and other butter substitutes.[47] Margarine was not an issue of major concern to the government. Sir John himself was just beginning to come back from serious attacks of bronchitis and sciatica which had kept him in bed for several weeks. William Fielding's secessionist activity in Nova Scotia, Honoré Mercier's nationalist movement in Quebec, the continuing unsettled conditions in the North West following the second rebellion and Louis Riel's execution, Ontario's bigoted response to the Riel uprising, the economic depression and faltering immigration, the fisheries' fight with the Americans—these were issues of much greater concern. But the parliamentary term of office was ending and Macdonald had to organize to fight the impending election. The government could not afford to ignore the pleas of the agrarian constituency, the most important segment of the electorate.

The initial government approach to the new threat to the butter industry was two-pronged: the levying of import duties and excise taxes, and the establishment of a licensing system for manufacture. In late March and early April, 1886, the Minister of Finance, Archibald McLelan, presented to the Commons a series of modifications to the customs and excise duties. A levy of 10¢ per pound on imported margarine, butterine, and all other butter substitutes was imposed. An excise tax of 8¢ per pound was to be applied on these items when manufactured in Canada.[48] In contrast, there had been an import duty of 4¢ a pound on butter since 1867; no excise tax had been levied on domestically produced butter.[49] McLelan's sole supporting argument was protection. "We hope not to receive any revenue from that source." That same day John Costigan also gave notice of motion: "That no oleomargarine or other substitutes for butter shall be manufactured except by persons duly licensed and that the Governor-in-Council may make regulations respecting such manufacture and the supervision thereof."[50]

The government's position, as enunciated by McLeland and Costigan and further developed in the debate by Mackenzie Bowell, Minister of Customs, placed margarine within the overall scheme of the National Policy. A high tariff would block importation, while competition of Canadian-made margarine with butter in the domestic market would be made very difficult by using the excise tax to fill the gap between costs of production. What was left was the opportunity of margarine being manufactured in Canada, as it had been for almost a year, but for export only.[51]

The running debate dealing with Taylor's bill and the several other government steps in April, May and June of 1886 is important because it was the first marshalling of arguments. Virtually all of the major arguments in the butter versus margarine controversy to be raised over the next ninety years were brought out, or at least touched upon. In fact, only one new argument was to emerge over the years, that of federal/provincial jurisdiction.

The men who spoke in this early debate set out several key points. Of the twenty-six speakers, four must be discounted. Costigan, McLelan and Bowell spoke because of their ministerial responsibilities. John Carling, a brewer from London, Ontario, spoke because he was Minister of Agriculture and was supporting his cabinet colleagues, as the brevity, thrust and timing of his own statement indicates. The backbenchers spoke because of some particular interest in the subject, but the issue did not split this group exactly along party lines. What is more significant is that a good number came from predominantly rural ridings with sizeable butter production, and located outside the larger urban centres.[52] (See Appendix 3.) All but three, Sydney Fisher from Quebec, Arthur Gillmor from New Brunswick and John Jenkins from Prince Edward Island, were from Ontario. What else could be expected? The shift from wheat to mixed farming was having the greatest impact in Ontario, as both Benjamin Allen from Ontario's Bruce Peninsula and James McMullen from central southern Ontario explicitly recognized. The subsistence agriculture of Quebec and the Maritimes was less specifically affected, except in pockets such as the ridings represented by Fisher, Gillmor and Jenkins.

The arguments put forth by the pro-margarine forces were to remain essentially the same over the decades, to be taken up by city-dwellers in the twentieth century. Properly manufactured and labelled, margarine could be a significant addition to the Canadian diet. More cheaply made than butter, it could serve the economic and nutritional needs of the poor. On the other hand, the anti-margarine arguments in the intensely emotional debate came from the agrarian sector, particularly in Ontario. The developing dairy industry argued that it and society needed protection from this competitor, and an outright ban, rather than permitting production under government regulation, was the preferred solution.

The problem of fraud in the food industry affected most of the industrial nations in the mid-nineteenth century. In the United States in particular, the writings of men such as Upton Sinclair, in *The Jungle*, provided evidence of corruption and extremely poor working and sanitary conditions in the industry.[53] Proposed remedies centred on the protection of producer and consumer. In Canada, George Taylor's resolution to regulate the manufacture and sale of margarine and other butter substitutes, by requiring their labelling with a name other than butter, was intended to protect the Canadian consumer from what Taylor called one of the "most glaring frauds ever perpetrated in this country."[54] Taylor congratulated the government for the proposed tariff adjustments, but believed these moves did not go far enough. Discussion with the Minister of Finance had elicited that there was no way of establishing how much butter substitute was coming into Canada, a statement corroborated by the Commissioner of Customs. The Commissioner believed that much was coming in under the guise of lard, grease and oil. The American and British experiences with fraud over the past ten years and their recent legislative actions to control the situation showed that Canada also needed to take steps. More strict American controls might even lead margarine manufacturers to "forcing the stuff into Canada."[55]

The concern over clean food reached a peak of emotional intensity quite late in the debate when Darbey Bergin, M.D. and Conservative member for Cornwall and Stormont, a riding in the heart of eastern Ontario's dairy district, delivered a stinging anti-margarine diatribe. He related some of the more revolting stories about American manufacturing processes; the gist of his comments was that the public needed to be protected from the deception which was possible when margarine and butter existed side by side in the marketplace.

> In fact oleomargarine is nothing but a highly purified kind of soap, and, as you perfume it, or do not perfume it, you have oleomargarine or scented soap. This compound of pork, as I told you, of diseased animals, is also made from the various fats that are picked from kitchens and gutters and the vilest purlieus of the different cities, and it is sold in almost every town in Canada as butter.[56]

A much more sensible position on the fraud issue had come from Joseph Jackson, a Liberal from South Norfolk and an operator of a lumber camp in Michigan, who had used margarine in feeding his workers. An important factor in the debate was that MPs were so unacquainted with the product that George Taylor had felt obliged to pass samples for tasting around among the members. Thus Jackson was the first speaker with practical experience with the food. He argued that Taylor's sample was a "very inferior" product. The margarine which Jackson had purchased for use in his lumber camp had met with no complaint from his men, while they objected to the varying tastes of the butter he had been using. Jackson gave no reply to Taylor's query whether the men knew they were eating margarine, a point which a later debater took as meaning that the men did not know. No one commented on the fact that since the men raised no complaints, they must have been satisfied with the product, at least in comparison with their previous complaints about lower grades of butter. Jackson continued by suggesting that the government should drop the idea of taxing margarine, so the poor could afford it more readily. The only requirement needed was proper labelling so that deception could not occur.

George Guillet (Conservative, West Northumberland, Ontario) noted a point to be developed more fully in the years to come. Distinctive colouring and packaging had been suggested to prevent fraud, but he opposed both as solutions because he believed that retailers could alter colouring and packaging to suit their purposes.

The need to protect producer and consumer from fraud emanated from the problems of quality control during manufacture and distribution. Bergin used another of his debating opportunities to describe the manufacturing process in the most revolting terms. He called for a government commission to check the New York manufacturing processes to establish the degree of purity of the final product. If there remained any doubt, Canadians and the dairy industry ought to be protected from the "vilest compound ever manufactured by chemical means." Bergin then explicitly contradicted himself by stating that the manufacturers "imitate what is called gilt-edge butter so closely that it is impossible for anyone to tell whether it is the genuine article or not," for it had the appearance, colour, and flavour of butter.[57] Samuel Hesson, a Conservative from Perth North, Ontario, raised another point by suggesting that much of the butter produced was "very good" but that poor handling, once it left the farm, spoiled an otherwise good product.[58]

Each food had its problems in the area of quality control, but margarine had a great cost advantage. In initiating debate Taylor had noted that margarine could be delivered into his riding at 14¢ a pound wholesale, a price with which butter had no chance to compete. (The Toronto *Globe* of April 16, 1886 noted butter wholesaling at 20¢ for the lowest grade, while Brockville creamery butter sold for the highest price of 28¢.) Differing estimates set the production costs for margarine from 9¢ to 14¢ per pound. The 10¢ import levy added to the 14¢ cost figure would place margarine at a price competitive with even the higher grades of butter. The best butter could stand the competition, for the difference in quality was appreciably in favour of butter. But second and third grades of butter stood to lose completely to margarine in the marketplace, for they were almost inedible, and would be purchased only by the poorest consumers.[59]

It is surprising that there was so little debate on the question of health, even though six of the twenty-six participants were doctors. This was probably due to the rudimentary state of knowledge regarding nutrition.

The power of the myth of milk in influencing the response to the arrival of margarine in the marketplace strengthened the economic argument. Dairying, and in particular butter, was too important to Canadian society to permit the potential damage that margarine could inflict. Taylor used the value of exports as the basis of comparison, and considered the category of "animals and their produce" as the most important in 1885, ahead of forest, grain, fish, minerals and manufactured products. He argued that dairying was the backbone of the animal husbandry category, with butter and cheese exports making up about one-third of the total. In noting the rumoured construction of a margarine factory in Montreal he did not judge it, as Bowell did, as part of a scheme to expand the Canadian industrial sector, but rather as all the more reason for immediate legislative action to protect the important Canadian dairy industry.[60]

The response of the Opposition showed at least some variety. Sydney Fisher, a Quebec dairy farmer and MP from Brome, a riding which in 1881 produced the second largest amount of butter in Canada, argued that he did not want the dairy industry protected at the expense of other sectors of the economy. He maintained that a healthful margarine, properly labelled, "will be merely an extra stimulus to them [the dairy industry] to make a better article with which this spurious article cannot compete." An improperly made or labelled margarine would be injurious to the interests of both the consumer and the dairy interests.[61] Later, Fisher would retreat from this position to argue that the government's tax measures would not lead to a cheaper food for the poor, so that he would support prohibition as the second-best alternative.[62] The leader of the Opposition, Edward Blake, representing West Durham, another rural Ontario riding, did not consider margarine a topic worthy of one of his famous lengthy discourses. In a brief statement, he supported his colleague Paterson's prohibition suggestion as much less complex than the government's proposed mode of action. Fisher's words summed up the Liberal position: "We are not protectionists, but we are prohibitionists."[63] In amplification of this particular point David Mills would respond, several weeks later, to Bowell's jibe that the Liberals' stand against margarine was not consistent with their vaunted free trade principles. He stated that his party had not arrived at its position on the basis of free trade or protection but, with "regard to a certain policy regulation . . . the protection of the public health."[64] Prohibition was in order.

The Commons debate of 1886 not only made clear the opinions of members of the House, and thereby the views of Canadian society, on the topic; it also gave impetus to the desire to impose a ban on margarine. The *Farmer's Advocate* had made first reference to the ban in 1884. In his introductory remarks Taylor had noted that the idea of total prohibition had already been supported by his constituents.[65] However, neither Taylor nor the government initially proposed such a drastic step. The first of three rounds of the debate saw the idea of a ban raised by Charles Hickey, M.D. from Conservative Dundas in the heart of eastern Ontario's dairy country — the area which produced the largest quantity of butter in Canada. It was best stated by George Casey, lawyer and member for West Elgin in southern Ontario. If, under the National Policy, mining and manufacturing were to be protected, then agriculture should be given the same treatment. Regulation was not enough, however. He feared fraud might occur even under Taylor's bill, as the manufacturer's proper labelling could be wrecked by the retailer "who would do his best to palm it off as genuine butter" when it really was "one of those horrible compounds of gelatine and petroleum and bristles and other things which the other gentleman described."[66]

In the second round of debate, the Ways and Means Committee's discussions of the proposed import duty of ten cents prompted Liberals William Paterson from South Brant, Ontario, and James Trow from South Perth, Ontario, to propose replacing the government's tax with outright prohibition of importation. Paterson argued that the only justification presented so far for margarine's existence was that it would provide a cheaper food for the poorer folk in Canada. An import duty would negate that goal by raising the cost of margarine to that of butter. The imposition of the duty would probably lead to a decline in margarine quality as inferior ingredients would be introduced to keep costs down.[67]

By the third round of debate all speakers were in support of a complete ban. The thrust was not against the manufacture or sale of a product which, it was admitted, could be made very nearly equal to butter in quality, but "against a base counterfeit of butter." Americans had attempted to use regulatory legislation to control margarine in their domestic market and had failed. It was contended that 1% of the margarine sold in the United States had been sold under that name; the rest was sold as "butter," creating much confusion. Several American states had recently turned to prohibition. The overseas markets for American butter had also been wrecked by margarine being fraudulently exported as butter. The same thing might very well happen to Canadian butter exports.

The 1886 debate led to a ban on margarine because myth and economics blended in support of primarily an economic consideration — the cost advantage which margarine would enjoy. Moreover, the fear that Canadian wholesalers and retailers would sell margarine as "butter," also made the extreme step of prohibition appear necessary.

Parliamentary action of 1878 to regulate margarine had raised no response in the contemporary press. The new debate drew at least some attention. In Prince Edward Island, neither of the Charlottetown papers noticed it, while in Nova Scotia and New Brunswick, coverage was limited to mentioning the various steps taken in the Commons, but with no editorial comment.[68] In Quebec, of the five papers examined, *L'Union des Cantons de L'Est* gave one brief notice of the ban.[69] The Stanstead *Journal*, with a constituency in the heart of Quebec dairying country, was the only paper to give extensive coverage. Drawing heavily on American experience with margarine, with its inadequate standards of cleanliness in manufacturing and the heavy substitution of margarine for butter, the *Journal* argued vehemently that an absolute ban was the only logical action Canadians could take. The protective impulse showed itself in the reasons expressed: to avoid unfair competition for the dairyman and deception of the consuming public.[70]

The most extensive coverage was found in Ontario. The seven papers surveyed reported on the debate in more or less detail. The Ottawa *Free Press*, a Liberal paper, published an editorial supporting the party's stand in favour of prohibition.[71] The Sarnia *Observer*, using American experience as the basis for its judgement, called on farmers to stand up and fight to keep "hog butter" out of the marketplace.[72] The *Monetary Times*, the business journal of

Toronto, limited its coverage to one paragraph expressing disappointment that the government had yielded to "unreasonable" pressure to shift to prohibition. "It [margarine] is as good as half the butter sold in Canada and should be allowed to be sold for what it is."[73] The Toronto *Globe*, the voice of Ontario Liberalism, covered the major steps of the debate in its "Parliamentary Summary" column. An article by Professor S. McBarre from the Ontario Agricultural College described the ingredients and procedures of manufacture and came to the conclusion that margarine should be prohibited in Canada.[74] The paper took the same position in four editorials. It also suggested, in true partisan spirit, that the Canadian farmers should thank the Liberal party for applying the pressure that forced the government to shift to the policy of prohibition. Justification was again protection: the dairy industry protected from competition, the consumer from fraud and from a threat to public health by a "compound of the most villainous character, which is often poisonous."[75]

West of the Great Lakes was a world that marched to its own tunes. Of the seven papers searched, only the *MacLeod Gazette* [Alberta] made any note and that was only one brief sentence regarding the government's intention to prohibit.[76] Of much greater concern to Westerners at this particular time was Irish Home Rule, the Métis uprising and local politics as reported in the *Manitoba Daily Free Press*. Even Jefferson Davis' latest speech took precedence. The conclusion has to be drawn that Ontarians were almost the sole Canadians concerned with margarine.

At the end of the first round of parliamentary debate the government had retreated. Finance Minister Archibald McLelan responded to the ban proposal by reaffirming his belief that the levying of the duties would serve the intent of prohibition, but that if the House were to prefer outright prohibition, he would be prepared to agree. John Costigan supported his colleague by expressing his belief that margarine could be "a really wholesome article of food" which could serve as a "really useful substitute for butter." An inspection system could keep out the poorer quality product. But he would agree to the alternative of direct prohibition if the House signified that this was its preference. The third minister to speak on the subject, Mackenzie Bowell, reiterated the government's original thrust: to use the two levies to permit manufacture of margarine in Canada, but for export only. A large quantity of margarine had already been manufactured in Montreal in the past year, out of pure tallow, and exported.[77]

At the end of the second round of the debate the government's position shifted to that of prohibition of importation, which the Paterson/Trow resolution espoused. Bowell proposed that Paterson drop his amendment and presented the government's own version. He argued that this approach would be simpler in that it moved directly to prohibition, while Paterson's called only for action and would require additional legislation to implement it. Paterson complied, noting he could still reintroduce his version if the suggested government action was not forthcoming. In anticipation of the Bowell resolution, the House then rejected the levying of a ten-cent duty on margarine imports.[78] The next day, action on the eight-cent excise tax on margarine and any other butter substitutes manufactured in Canada was also postponed.[79]

Having effectively killed the government's initial proposal for dealing with butter substitutes, the Commons' next move came three weeks later when it resumed debate on Taylor's motion to regulate. The preliminary thrust of this round centred on the shifts in tactics by Taylor and the government. Since his initial step, Taylor had taken note of what he saw as a swelling of opposition to margarine becoming evident in newspapers, in resolutions from dairymen's associations, and in correspondence to him from across the country. In response, he now gave notice that he intended to replace his original proposal to regulate with a bill that would prohibit manufacture and use of margarine, the import ban having already been promised by the government.[80]

After some wrangling, the House voted to adjourn debate on Taylor's resolution. It was understood, although somewhat tenuously, that the government would propose a total ban, even though Costigan's bill to establish a licensing system for manufacture was still on the order paper.[81] The following day Hector Langevin, Minister of Public Works and second-in-command in the ministry, firmly committed it to prohibition when the House, in committee of the whole, dealt with section 12 of the amendments to the Inland Revenue Act. In view of the recent action, Costigan now proposed deleting the section dealing with controlling the manufacture of margarine.[82]

Two more weeks passed before the House, in committee, came to move Resolution 17 which called for the prohibition of the importation and manufacture of margarine and all other butter substitutes. Then on June 1, McLelan presented the bill based on the domestic aspects of Resolution 17, concerned with manufacture and sale, which was quickly given three readings and sent to the Senate.[83] The Senate in turn gave its consent without real debate[84] and royal assent was given later the same day. The final procedure was to pass the Customs Duties Act amendment[85] by which margarine, butterine, and all other substitutes for butter were added to the list of articles, the importation of which was prohibited.[86]

The extended debate and manoeuvring had ended. Passage of "An Act to Prohibit the Manufacture and Sale of Certain Substitutes for Butter" outlawed margarine for the next sixty years, except for a brief interlude, 1917-24, when wartime pressures demanded use of all available food resources. The preamble, which stipulated that action was required only because use was "injurious to health," certainly did not offer the most important reason, as an analysis of the debate demonstrates. The process leading to a "bad" product, not the product *per se*, was the real villain. Hence, the legislation, in order to justify the radical step of prohibition, had to present it in as bad a light as possible.

The margarine debate did not provide the basis for any examination of the fundamentals underlying the Canadian national identity. In contrast, the contemporaneous American debate did provide the occasion for congressional participants to touch on bedrock once again.[87]

A rural society had fought to protect itself. Margarine had arrived at a time when many Ontario farmers perceived themselves to be extremely vulnerable because of the recent shift from wheat to diversified agriculture. They were also insecure as to their future livelihood. If the agrarian-myth factor is added to the great economic importance of milk and butter, then the vehemence with which nineteenth-century Canadians opposed the use of margarine can be readily understood. Ontario dairymen chose the ban, and cabinet and parliament listened to this major segment of their constituents. Canada got what Ontario wanted.

CHAPTER TWO

Exclusion
1886-1914

> Ministers of Agriculture, Dominion Commissioners, Tories, Grits, Patrons, Yankees, Senators, Knights, Members of Parliament, Ministers of the Gospel, Doctors and Citizens generally were all one in their allegiance and loyalty to the kingdom of the cow.[1]

During the quarter century following 1886 the butter industry strengthened its position within the Canadian economy. It used this new power to insist upon a tightening of the ban against margarine. Although this period was to see the influence of the butter lobby at its peak, it was also the time when the first major crack in its armour appeared. The population flood of the Laurier era dramatically increased the domestic market for all milk products to the point at which the demand for butter exports could no longer be met. When the effects of war on the food supply were added, the return of margarine upon the scene was highly likely.

The wheat-to-mixed-speciality farming revolution that had begun in the pre-Confederation era continued, its course mirrored in the expanding number of milk cows.[2] In 1871, that number stood at about one-and-one-quarter million. Over half were found in Ontario, another third in Quebec, some one-tenth in Nova Scotia and even fewer in New Brunswick. The national total grew over the next five decades by a dramatic one-and-a-half million.

The greatest period of increase occurred in the 1890s when over half a million head were added to the national herd. Such growth did not affect the Maritimes to the same degree as the rest of the provinces, and ended for them by 1880. For the central provinces the halt came twenty years later. The expansive wave affected the Western provinces the longest, quite in line with the dramatic growth of settlement in the Laurier years. (See Appendix 4.)

Not only were more cows raised, but the quality of stock was upgraded significantly. Between 1910 and 1920, the average annual milk yield per cow rose by 6%. More and better cows and a doubling of milk prices laid the basis for a strong, prosperous dairy industry.[3] This success showed itself in the virtual doubling of total milk production in Canada during these years.[4]

Optimum growth, however, met head-on with the harsh realities of the Canadian environment. Climatic conditions, in particular the severity of winter, determined that even where dairying was possible, cows were milked only during the spring-summer-autumn seasons when pasture was available. For more than half the year, most Canadian cows needed supplemental feeding. In the butter industry, this pattern showed itself in a feast/famine production cycle. Dairymen attempted to meet the problem and to provide themselves with a steady income by holding back some of the butter produced in summer for fall and winter sales. But this resulted in stale butter, which in turn influenced sales and prices. Moreover, each change of feed — grass in the spring, dry feed in the fall — also affected the colour and flavour of the butter.

The effects of inclement weather could be devastating. For example, the winter of 1906-07 was extremely severe in Alberta. The impact was compounded by the long delay in sufficient grass growth for pasture in the spring. Loss of cattle ranged from 5% to 90%, as feed and water, even where available, could often not be delivered to isolated herds. The calf herd of that spring was also severely hit. In both the short and the long term, milk production and thus that of butter and cheese was severely reduced, as herds shrank in quantity and quality.[5]

There was still much lacking in the quality of Canadian butter. At the World's Columbian Exposition in Chicago, in 1892, Canadian cheese producers' 22,000-pound cheese had attracted much attention.[6] Their 849 entries had won 736 medals and diplomas, of which 136 were firsts. In contrast Canadian butter had done miserably. Of 207 exhibits, only 40 had been awarded medals or diplomas of any kind: none had achieved the lofty level of a first[7] and adulteration was prevalent.[8] The reason for this lack of success may well have been the greater tendency of dairymen to produce butter through the casual labour of the family on the farm, with a lack of stress on improving quality.

Although adequate winter stabling was an urgent need, the concern for proper feed was even greater. Some work was done to improve pasture for the productive season. For example, winter rye was planted in Prince Edward Island. More emphasis was placed upon the provision of adequate quantities of hay and ensilage, primarily fodder corn. Winter carry-over of a herd became possible. Even more importantly, the quality of feed was upgraded, so that the cows sustained sufficient milk production to make possible year-round butter manufacture.[9]

Another innovation was the trend towards fitting out cheese factories for the production of butter in the October to May off-season. By 1894, this procedure was well-established in Ontario and well-introduced in the other provinces. In those regions where butter creameries were not yet fully established, it provided the dairy farmers with more revenue, and as a result more and better cows could be kept. The use of the butter/milk by-products to feed hogs gave even more income.[10] The capability to keep herds and facilities in year-round production was one of the greatest advances for the Canadian butter industry in its worldwide competition with, for instance, the industries of Denmark and New Zealand, where climatic conditions were much more favourable.

Canadian dairy experts continued to urge the use of creameries to manufacture butter, in order to achieve the required standards of quality. Proper methods of production through the whole system meant the most profit. It was observed that, for the Alberta dairymen who had been properly equipped and had the supplies to care for their stock, the ravages of the severe winter of 1906-07 had been relatively light. Where they had also used creameries to produce the best butter possible, the higher prices in a butter-short market had led to increased profits.[11] The evidence was available: the problem was to get Canadian farmers to apply it for their own gain.

The other major area of activity aimed at upgrading the butter industry was that of transport. In spite of the fact that the introduction of the cream separator in the 1880s cut the weight to be hauled to the creamery by the farmer by 90%,[12] the conditions of many of the rural roads remained a serious obstacle. As late as 1895, it was said that for Ontario—and probably typical for the rest of Canada—"road making is not only an important question, it is *the* question of the day."[13] Rebuilding Ontario roads to British standards would cut the cost of hauling milk by half and save Ontario farmers $400,000 annually. The demand for better roads was to remain constant for a long time to come.

Adequate transport from creamery to market was equally vital. Trains, with their ability to move bulk year-round, made a great difference. In the export field, the distance to the British market gave cheese, with better keeping qualities, an early advantage over butter. In 1895, the federal government responded to the butter interests' representations in two ways. At various centres in Ontario and Quebec butter came to be stored under refrigeration. Refrigerated railway cars were attached to trains running to Montreal. Shippers were charged the normal rates: the government paid the difference. Secondly, the federal government, on the basis of earlier tests, contracted for the building of several ships with refrigerated sections for the storage of butter. A third transport innovation, initiated by government in 1891, was shipping increasing amounts of winter-made butter to Britain during the winter, rather than saving it for later shipment. The quality of the fresher butter was such as to attract an additional 3 to 4¢ per pound.[14]

The butter industry used government to protect its interests in several other ways. In 1889, the first national dairy commissioner was appointed to lead the government's work in support of all dairy activities.[15] Between 1891 and 1893, branch experimental stations were set up in each of the five Maritime and central provinces. The prime purpose of the stations was to educate the local farmers as to the viability of year-round dairying.[16] The department also used its personnel to teach the dairymen the value of keeping records on the quantity of milk and of butterfat produced by each cow.[17] The federal government, and at least Alberta and Saskatchewan on the provincial level, became directly involved in the operation of creameries. This paralleled the development of cooperatively owned creameries.[18]

These technological innovations were part of the great expansion of the milk-supply base, and hence of butter production. On the demand side the chief factors were the insatiable British market for butter, and the continued expansion of Canada's population and thus of butter consumption.

The first use made of the additional supply of butter was an attempt to expand the Canadian share of the British butter market. This market, consistently to 1913, took the vast majority of Canadian butter, even though the Canadian share was minuscule. Beginning in 1858, Canadian exports held relatively steady until 1883, when, within one year, they dropped 47%. After six years a dramatic upswing occurred, doubling exports in each of the next two years.[19]

In mid-century, the Americans had come to need their dairy products to meet their own food requirements, enabling Canadian cheese to dominate the British market until World War I.[20] On the other hand, Canadian butter did not create for itself as large or as secure a niche (see Appendix 5). There were steady complaints from Britain concerning quality. For Canada the British market was vital: the United States tariff wall was increasingly hard to jump, and in 1890 the McKinley tariff closed this avenue completely.

Growth in the domestic market was generated by the increase in the number of consumers and in the rate of per-capita consumption of milk products. In the latter years of the nineteenth century the steady expansion of Canada's population generated by the natural birthrate between 1871 and World War I had usually been offset by the net loss in immigration. With the turn of the century the flood of immigrants significantly intensified the population growth pattern. The domestic market grew to the extent that Canada could consistently consume all that its butter industry could produce. When statistical evidence is combined with the data on urban growth,[21] the main reasons for the 400% increase in the growth of the Canadian market for milk products becomes evident. The urban growth rate reached its peak of 62% for one decade between 1901 and 1911.[22] Domestic consumption of all dairy products also increased: 75% between 1904 and 1914. Per capita consumption of milk and traditional dairy products such as butter, cheese and cream, plus the new lines such as condensed milk, powdered milk and ice cream, increased 32%. A 43% increase in production could not meet the pressures generated by this expanding consumption. So, whereas in 1903 exports claimed 40% of total Canadian butter production, only 20% was being exported by 1913.[23] In the 1912-13 production year, the most notable feature of the export trade was the fact that for the first time in over sixty years no butter was shipped to Britain.[24]

The massive demographic expansion of the urban sector of Canada stimulated the productivity of the dairy industry. This growth had wider implications. For rural Canadians transplanted to the cities, the "good old days" became those of the past, "down on the farm." Thus the power of the agrar-

Table 2-1
Population Increase Re Urbanization

	Total Population	Urban Population	Urban Population Percent of Total
1871	3,689,000	859,537	18.3
1921	8,788,000	4,165,512	47.4
Increase	5,099,000	3,305,975	

ian myth was greatly increased, carrying with it the myth of milk as the "perfect food." In the era before the structure of food was fully understood, myth often provided the medium for an understanding of nutrition. It was not until 1901 to 1905 that a scientific knowledge of vitamins and their role in nutrition was added to that of the other groups of food essentials: proteins, fats, carbohydrates and minerals. Even then it was not until after World War I that this new information began to affect medical and nutritional practices.[25] Penetration of this knowledge into the cultural milieu would take even longer.

In the nineteenth century conception of these matters there were two types of food: that which replenished energy and that which, in addition, provided the body with the capability for new growth. All foods could serve the first purpose, but foods capable of serving the second were very rare indeed. Milk and its by-products were the foods most often perceived as fulfilling that function. Since children were in the process of building their bodies, it was absolutely essential that they be given the foods capable of generating that growth. Within this context the dairy industry was portrayed in idyllic terms:

> In the dairy is the occurrence of daily miracle — the transmitting of golden sunshine, through the blossom and the grass, into golden butter — and in this transmission is involved all the mysterious, subtle forces in the air above, the earth beneath, and the waters under the earth, whose sum total we call "nature" and whose understanding "science."[26]

The producer of this elixir was worshipped in the same journal as "Nature's own symbol: she is the Greek Astarte and the Syrian Ashtaroth, the Babylonian Mylitta, and the Egyptian Osiris and Isis."[27]

The most fully developed statement of the agrarian myth of milk was published in the *Maritime Farmer* during World War I, as the butter industry did its utmost to educate the public to the wonders of its product:

> Therefore, butter is an article of food, not only capable of maintaining animal life, but of promoting the growth of animal life.... When you buy oleo, you purchase animal and vegetable fats in various proportions, that have more or less proven to maintain life, but when you buy butter you have this same power plus the power to promote growth in the human race. You get something when you buy butter that you do not get when

you buy oleomargarine and it is this quality of promoting growth that has given butter its universal use among enlightened and intelligent people, and give[n] unto milk the place it now occupies.[28]

The second (the first was beet sugar for cane sugar) of the substitutes in the history of food,[29] margarine attacked the holy of holies and generated a powerful, over-protective reaction.

Because advances in margarine technology took place outside of Canada, only brief notice of them is needed here. There were three major advances in margarine technology: (1) the use of cultures of lacto bacteria to give the flavour of butter; (2) the introduction of vegetable oils and fats to produce the so-called nuts and milk margarines; and (3) the introduction of hydrogenated oils. The evidence which occasionally surfaces suggests that those Canadians who were interested in this product paid attention — but they were a small minority.

These changes merely provided more reason for the opponents of margarine to promote their concerns. For example, the hydrogenation process permitted the use of a wider range of cheaper vegetable and marine oils, but contained few vitamins. The scientific understanding of vitamins led to the recognition that animal fats, such as the beef suet used earlier in margarine, contained far fewer vitamins than milk — incorrectly — was perceived to have. The use of some milk in the manufacture of margarine could not bring it up to the level of quality of butter, at least in the eyes of the dairy lobby.

In December 1913, the *Farmer's Advocate* noted that the pressures to legalize margarine were "great." Others were arguing that the farmer would gain from the increased price for cattle, hogs, packing-house scraps, and fats. However, the gist of the article was to urge that Canada stay out of the food substitutes field. An army of inspectors could not keep out the fraudulent products that would result. Consumers might pay less for "honest butter" but would pay more for meats. Only the "big interests" would gain:[30] margarine was the product of the small number of large food-manufacturing and packing companies, the "bad guys" of the agrarian myth, while butter was the product of the work of the multitude of independent farmers, the "good guys." The margarine industry was "parasitic,"[31] living off the products discarded by the dairy industry, producer of the perfect food.

Still, there were differences of opinion within the industry. In response to the *Advocate's* propaganda effort, a Nova Scotia dairy farmer felt strongly enough to object to the *Advocate's* support of the new dairy bill before the Commons in April, 1914. In a letter to the editor, he put out the "class" argument: margarine would help the poor who could not afford the more expensive butter. Instead of excluding all butter substitutes, proper government regulation to control fraud was all that was needed.[32]

Reference to the Dairy Act of 1914 brings the focus back to the legislative field which the butter interests used to protect their position. The ultimate goal of the new round of legislation was to clamp down even more tightly on all dairy substitutes, especially with respect to butter and cheese. The idea was to deal not only with "potential" threats, but with "real" ones. This could best be done by preventing the establishment of plants for the production of any sort of substitute in Canada.[33]

The Inspection and Sales Act of 1896 was a consolidation of the several dairy acts passed since the ban on margarine had first been adopted in 1886. A housekeeping piece of legislation, section 298 of part VIII dealt with margarine and other butter substitutes. In 1903, the Dairy Act did make a substantive change in the legislation. Ironically, it was introduced by Sydney Fisher who had proposed a sensible way to handle the butter/margarine problem in the initial major debate seventeen years earlier. Now, however, as Laurier's Minister of Agriculture, he proposed even stronger regulations to protect butter. The earlier negative reaction had come as a result of the imminent arrival of margarine in Canada. This time, the product threatening the protected position of butter was process butter.[34]

In introducing the legislation, Fisher admitted that the main purpose was to protect the interests of the Canadian dairy industry. Bill 207 forbade the manufacture of process or of renovated butter;[35] it required butter to be properly labelled; corrected the vagueness of earlier legislation with a stipulation that butter with more than 16% water content was to be ruled adulterated; required that no chemical could be added to butter to make it absorb water; and strengthened penalties for violations. Fisher argued that the 1886 ban on margarine had performed its task of keeping the Canadian market "pure" for butter and that this additional legislation was required now that a new threat had appeared.[36]

The question, as in 1886, was how harshly to deal with butter substitutes. The debate did not break down on party lines, and once more was led by the members from rural Ontario. A factory in Manitoba was manufacturing process butter, and this threat needed to be dealt with. Beyond the twitting of members on their change of positions since 1886 (generated to a large degree by the 1896 switch of parties in power), the major question was whether the 1886 legislation did not already cover the new situation. Fisher pointed out that the existing laws banned butter substitutes "manufactured from any animal substance other than milk." Since process butter was derived from milk, the prohibitory net needed to be cast wider. Again, the American experience was called upon to strengthen the rationale for Canadian action. On the positive side, the intent of the new legislation was to lay out better ground rules to distinguish creamery butter from dairy or farm-made butter, and thus strengthen the position of Canadian creamery butter in the British market.[37]

The 1903 Dairy Act prohibited the manufacture or sale of all dairy substitutes. But an underground market for butter substitutes indeed existed, as demonstrated by the two separate legal actions against producers of such products in 1908.[38] The potential "hole in the dike" presented by the reciprocity negotiations in 1911 brought into question whether margarine would be permitted under the new tariff arrangements. The answer was an unequivocal "no."[39]

The eve of World War I set the stage for the next episode. The specific problem was that a judge had recently ruled that there was no proper authority to convict persons charged under the 1896 Inspection and Sales Act. In Montreal, in particular, people were inserting water into butter up to levels of 40% or 50%. Bill 112, by selecting the relevant sections from the Inspection and Sales Act and setting up a new Dairy Act, was intended to tighten the law.

Several attempts at common sense were made during the debate. For example, E.W. Nesbitt, from North Oxford in southern Ontario, drew the attention of the House to the fact that margarine manufactured from beef suet kept as well as butter did and that it was as "pure and wholesome" as butter. It should be made available for the poor, but it did require proper labelling.[40] Moreover, the Senate did more than "rubber stamp" the new bill. The point is worth noting because it was the Senate which, thirty-five years later, was to be the focus of the successful effort to repeal the ban. Several senators raised the needs of the poor as a major reason for legalization. Butter, selling at between 40¢ and 50¢ per pound, was beyond the reach of a married labourer, with ten children, earning $1.50 an hour. As in many such instances, the debater probably found the most extreme example to drive home his point, but the point was there to be made.

Senator Lawrence Power of Nova Scotia argued two points that would weigh heavily against butter in the final battle of the 1940s. Whereas the initial ban had been passed to help develop the export trade in butter and cheese, this help could no longer be expected nor afforded. Canada was now an importer of butter. The prohibitory law was, therefore, not aiding Canadian dairymen, but rather those of Australia and New Zealand. His second point was one which Senator W.D. Euler was to stress heavily when he took up the cudgel. "The law is an improper infringement upon the liberty of the subject." The British example of allowing sale but penalizing fraud ought to be copied.[41]

In the end, the need to protect butter from competition, albeit disguised as fear of fraudulent competitive products, won out. That spokesman of the farmers' interests, the *Farmer's Advocate*, reminded its readers that the battle for the kingdom of the cow was not yet won, and that margarine interests were agitating to have its manufacture and sale made legal in Canada. What these "interests" were was not set out, but they had to be the packinghouses that were to manufacture the product when the ban was lifted in 1917. The *Advocate* gave its blessing to this latest protective effort by the Borden government, since the dairy industry "would now be on as sound a basis as any legislation could put it."[42]

Before World War I erupted, circumstances already had begun to challenge the basis upon which the butter interests had made their case for protection. The war was to intensify these pressures to the point that the ban was to be lifted for a short time.

CHAPTER THREE

Momentary Easing 1914-19

The population increase during the Laurier era was of such magnitude that the butter industry could no longer meet the needs of the Canadian market. The special pressures generated by World War I further intensified the demand for butter—for the home front, the battle front and for Canada's allies. All of these needs could not be met, even with substantial increases in production. Despite objections by the butter lobby, in 1917 margarine was legalized as a temporary wartime measure. Although the ban was reimposed soon after the war, this initial break, and the arguments used by both sides, demonstrated that the rationale underlying the imposition of the ban in 1886 was no longer viable. The debate also showed that the science of nutrition had developed to the point at which the place of butter in the food spectrum—and that of margarine—could be properly appreciated. Milk was demythologized.

In the decades after Confederation, one of the main questions facing Canadians was to what degree and on what grounds they would become involved in international affairs. World crises, wars and threats of wars in particular, served as occasions when increased autonomy was sought in order to achieve responses which would be properly "Canadian." Few Canadians objected to Laurier's 1910 statement: "When Britain is at war, Canada is at war." However, the degree and manner of participation became hotly debated issues. On the one side were those who, like Robert Borden, argued vigorously that Canada could not afford to hold back on any contribution to ensure victory. On the other side were Sir Wilfrid Laurier and those who contended that special Canadian circumstances imposed limits on the nation's contributions to the war effort. In any case, the war would be decided by the actions of the major powers; the Canadian contributions could never be decisive. The issue was most significantly argued in the area of manpower, culminating in the conscription crisis of 1917.

In many ways food was just as important. In April 1917, the *Halifax Herald* carried on its editorial page a cartoon showing a witch-like figure labelled "Hunger" holding the world in its hand. The caption read: "Bread More Essential Than Bullets." The previous Sunday an appeal for greater food production had been made in the various churches in Halifax.[1] Canada's war contribution in the form of various foods such as bacon, butter, cheese and wheat was more vital to victory than her soldiers or manufactured war material.

Up to 1916 the war effort was generated by volunteerism, nurtured by patriotism — a relatively low-key phase. Food production followed the paths of encouragement, restriction, and persuasion, to conserve on the home front and to increase production so that greater quantities could be exported. Butter was one of the important food products in the Canadian production and conservation schemes. The offspring of the "perfect food" had been a major export item for many years, and the loss of self-sufficiency in butter production had come almost coincidentally with the outbreak of the war. That all-important fact had not yet had time to permeate popular thinking.

The volunteerist efforts of the 1914-16 half of the war were enough to meet the food requirements on the home and war fronts. What rumblings there were concerning the legalization of margarine were not enough to move the editors of the *Farmer's Advocate* to raise the spectre of the margarine "menace" until early November, 1916.[2] In 1915, a good season for feed grains kept the price of butter close to average over the winter. But 1916, the worst weather ever experienced over most of Canada, led to a drastic drop in milk production. The resulting decline in butter supplies, combined with the renewal of heavy exports, led inevitably to rising prices. The average wholesale price reached 30.07¢ per pound, up from 27.60¢ in 1915.[3]

The pressures on butter supplies led to an intensification of agitation to have the sale of margarine legalized.[4] In response, the anti-margarine struggle re-emerged, first of all on the government inspection front. The annual report for 1916-17 by the federal Minister of Agriculture noted two convictions had been secured for the selling of butter substitutes,[5] although the type was not specified. Between January and March, 1917, federal inspectors, in response to many complaints concerning extensive sales of margarine, checked 228 samples of butter. Not a single sample of margarine was found.[6] Two convictions, the first since pre-war years, suggested some increase in illegal trafficking but surely no widespread black market.

In the last two years of the war the question of survival, let alone victory, came to require tighter control of all aspects of the national war effort. When regulation was emphasized, food control procedures were a mixture of British compulsion and American volunteerism, with the stress on the latter.[7] W.J. Hanna's appointment as Food Controller, in 1917, was followed by the establishment, in 1918, of the Canada Food Board, headed by Thomas Crerar, Minister of Agriculture.

The first shot in the renewed conflict between the two foods was fired January 17, 1917. The Saskatchewan Dairyman's Association responded to pro-margarine rumblings by passing a resolution opposing legalization.[8] This meeting led also to correspondence between the various butter interests inaugurating a more concerted effort to defeat this latest threat. The suggestion was to set up a central committee to coordinate the fight and to collect money to pay for the struggle. The correspondents illustrated the spectrum of members of the butter lobby.[9]

The anti-margarine agitation did not have deep roots, either in the agricultural community or in Canadian society at large. It never generated the large number of petitions to parliament brought forth by the liquor prohibition issue, nor even the much smaller number engendered by the conscription issue. Only one petition against margarine, signed by 29,000 Quebec farmers, was laid before parliament;[10] protests raised by the United Farmers of British Columbia[11] apparently were not forwarded to Ottawa. The Cattle Breeders' Associations sent a deputation to the Minister of Agriculture protesting margarine's use.[12] Illustrative of the narrowness of the effort against margarine is the fate of an anti-margarine resolution submitted to the annual meeting of the United Farmers of Alberta. No one appeared in its support, and it was "not dealt with."[13] An effort to win support from the Saskatchewan Grain Growers' Association resulted in a tabled motion.[14] The dairymen in Manitoba, Saskatchewan and Alberta succeeded in running anti-margarine resolutions through their organizations.[15] The lack of a common ground among Canadian farmers was to plague the butter interests in their approaching fight.

The agitation against margarine concentrated in the dairy industry: dairymen, creamery men, and their suppliers. With W.A. Wilson, Dairy Commissioner for Saskatchewan, as the coordinator, these forces concentrated their efforts on bringing pressure on the federal government, and in particular the Minister of Agriculture.[16] They were joined by the editors of *Farmer's Advocate*, the longstanding opponent of margarine.

In their own defence, the dairy interests once more rehearsed their basic line of argument. Dairy farmers, suffering the effects of labour shortages, high wages and expensive feed, viewed the added burden of competition from margarine as the last straw which would force many into other lines of work. Secondly, consumer interests would not be advanced by margarine's presence in the marketplace. Although the packing houses might set initial low prices for margarine, this would only be a tactic aimed at driving the dairy farmer out of business, after which prices could be raised. The American experience was that margarine sold at only 4 to 5¢ per pound less than butter, even though production costs were about half those of butter. The poor people would gain no advantage unless the Food Controller set the price of margarine to meet their needs. Thirdly, the familiar warnings concerning fraud were resurrected. One refinement was added: if it was to be sold, margarine would have to be uncoloured, as was the rule in Denmark. The assumption was that because Canadians were conditioned to expect their bread spread to be coloured yellow, they would be less likely to purchase an uncoloured product. Lastly, the packing houses, one of the traditional enemies of the farmer, would make the product; meaning, by definition, that the farmer's interests would not be served.[17]

Remarkably, no effort was made at this time to raise the argument which had been used in 1886 as the prime reason for introducing the ban — protection of consumer health. This was because, by 1917, the idea that margarine might not be healthful no longer carried any weight in the light of the experience of other countries.

In Canada, the first round in support of the margarine-is-healthy contention came from the Canadian Medical Association. In April 1917, an editorial in the CMA *Journal* categorically supported the consumption of margarine as equally nutritious as butter, less likely to become rancid, and significantly less expensive.[18] No evidence has been found of contemporary newspaper coverage of this statement, and it is doubtful whether this judgement filtered to the general public through the doctors' counselling of patients to any great degree.

The second missile came from a source even less influential. The science of nutrition had progressed to the point at which reasonably sound analyses could be made of the nutritional value of each food and of the total food values required by persons of differing body structure and occupational activity. In the summer of 1917, the federal government issued a bulletin aimed at providing nutritional information for Canadians who were having difficulty in coping with the inflationary price spiral. The document can be taken as presenting, in layman's terminology, the latest understanding of the structure of food and of human dietary needs prepared by experts in the field. Here was evidence enough to demythologize milk and butter and to put them in their proper place in the food spectrum. Here also was evidence to show that margarine was a healthful food, even more healthful, in fact, than butter.

The two nineteenth-century conceptions of the nutritional purposes of food remained: "to build up the material of the body and to replace tissue which is wasted in life processes," [proteids or protein] and "to furnish energy or the power to do work" [fats and carbohydrates]. The unit of energy had been labelled "calorie." It was recognized that the various foods might contain, in varying proportions, all three types of food. It was argued that the most important factor for the consumer to remember was that protein, per ounce, was capable of producing 116 units of energy or calories, fats 264, and carbohydrates 116. The average male required 2810 calories per day, the average woman 2240. The ratio of proteins:fats:carbohydrates for the man should be 4.162:1.975:16.633. For the woman the ratio should be 3.316:1.587:14.110. The table below was provided to guide people as to typical personal needs, and to give examples of the daily quantity of food needed by specific individuals and the calculated energy represented by this food. Proteins could not be replaced by fats or carbohydrates but the latter two classes could, to a large extent, be interchanged even though fats were considered more important.

There were two major weaknesses in this analysis. First, there was an almost total failure to recognize the role of minerals. They were stated to be

Table 3-1
Recommended Daily Food Needs (1917)
(quantities in ounces avoirdupois; energy in calories)

Description	Proteids	Fats	Carbohydrates	Calories
Soldier during peace	4.197	1.411	18.659	2784
Soldier in light service	4.127	1.235	15.768	2424
Soldier in the field	5.150	1.623	17.778	2852
Labourer at work	4.586	1.411	19.400	2903
Labourer at rest	4.832	2.510	12.416	2458
Cabinet maker (40 years)	4.621	2.398	17.425	2835
Young physician	4.480	3.171	10.143	2602
Labourer	4.691	3.351	14.885	2902
English smith	6.208	2.505	23.492	3780
English pugilist	10.159	3.104	3.280	2189
Bavarian woodman	4.762	7.337	30.900	5589
Silesian labourer	2.822	0.564	19.171	2518
Seamstress in London	1.904	1.023	10.300	1688
Swedish labourer	4.727	2.787	17.108	3019
Japanese student	2.928	0.494	21.941	2779
Japanese shopman	1.940	0.212	13.898	1744
Eskimo (Krough)	9.947	1.443	1.792	2604
Bengali	1.834	0.907	16.649	2390

of "equal importance," yet no calculation of their content in food, or of the body's requirements, was presented. "They [are] ordinarily present in our foods, and we take them incidentally and of necessity when we use natural food materials." The second weakness lay in the inadequate appreciation of the role of vitamins in the diet. All that was said was that vitamins (there was no recognition of the different ones) were "always present" in natural foods. They "greatly influence the efficiency of food, as regards growth, and the prevention and cure of disease, although in amounts so small as to have escaped attention until recently." As a result of these failures to perceive the proper role of minerals and vitamins, fruits and vegetables were considered as having an "apparently negligible value as nutrients." Vegetables were valuable for the bulk they provided, but nothing was said regarding the value of fruits.

A list of 131 commonly eaten foods was presented, with a breakdown of protein, fats and carbohydrates contained, together with the cost of one pound of each in Ottawa as of June 1917. One pound of butter contained 0.122 ounces of protein, 13.392 ounces of fats, and 0.080 ounces of carbohydrates. Margarine made the list, even though its presence in Canada was not yet legal. It contained 0.158 ounces of protein, 14.014 ounces of fats, but no carbohydrates.[19]

Given the ranking of the three food classes set forth in this bulletin, margarine had to be recognized in 1917 as not only healthful but as having greater food value than butter. The important fact that butter did, and margarine did not, contain generous amounts of vitamins A and D, both vital for body growth and functioning, was then unknown. Any Canadian who in 1917 judged butter to be of greater nutritional value than margarine, did so strictly on irrational grounds — a "gut reaction" conditioned by the prevailing myths. Only when vitamins came to be better understood was butter placed ahead of margarine, ensuring that margarine manufacturers had to catch up by inserting artificial vitamins.

On the basis of such information, limited though it may have been, other groups joined the fray. In Ottawa, the Ottawa Anti-Tuberculosis Association passed a resolution calling for an end to the ban on margarine. The action was sent to government officials and received some newspaper coverage. It argued that there was an "alarming increase" in the incidence of tuberculosis in Europe, caused for the most part by a lack of nourishing food. Lest the same fate befall Canadians, the ban on margarine needed to be removed as part of doing everything possible to provide them with adequate food and thus good health.[20]

In lifting the ban, the politicians' chief consideration was identical to that which supposedly had inspired the 1886 prohibition — the protection of public health. There was now, however, an ironic twist: the protection of that same public's health now required the legalization of margarine sales.

The activities and arguments in support of margarine show the complexity of the issue as well as the growth of that support. The CMA position was really expressing what a large segment of the general public already accepted. Three months before the medical journal's editorial was published, a farmer's daughter had written to the *Farmer's Advocate*, supporting margarine as a "perfectly clean food product." She argued that farmers were too broad-minded to desire continuation of the ban as purely class legislation: margarine would be "welcomed as an adequate and cheap substitute" by those unable to afford butter.[21] In the same issue, a farmer also supported this "wholesome and cheap article." He suggested margarine be coloured, like maple sugar, to prevent fraud.[22] Among the letters to the editor of the *Advocate* such letters outnumbered those of the proponents of butter. On another front, the Labour Council of Montreal sent to its city council, for forwarding to the federal government, a resolution in favour of the manufacture and sale of margarine.[23]

The correspondence of the men leading the butter lobby's fight with members of parliament illustrates the lack of clear-cut options for the MPs in deciding how to deal with margarine. J.G. Turrell noted that American dairy farmers were handling the competition. He recognized that there was a significant price differential in Britain; thus, margarine was probably a real boon to the poor. Margarine was "a very good substitute" for butter, but he would wait to hear what his Saskatchewan farm constituents had to say before making up his mind how to vote.[24] D.B. Nealy, another Saskatchewan MP, would also wait before reaching a decision, but in the meantime he had difficulty reconciling the anti-margarine position with the "well-known free trade principles" of the prairie farmer.²⁵ The dilemma was to become increasingly evident. It was to show itself very strongly in the post-war debate over reimposition of the ban. Even the butter industry's allies, the creamery operators, could not view the issue without some bewilderment as to which side to take. It was accepted that margarine in the marketplace would force a decline in the price of milk by-products, such as butter. But it could also be argued that a benefit would accrue to the creameries. Margarine was of better quality than the poorer grades of butter, in particular the farm-made product. Thus, margarine's competition might force many farmers to stop making butter on the farm, and to ship all their cream to a creamery for manufacture into best quality butter.²⁶

This mix of arguments was not decisive. What was conclusive was that in the context of a growing world crisis in food supplies which threatened famine,²⁷ the butter industry, in spite of an all-out effort to expand production as part of Canada's contribution to the war effort, could not meet the continuing requirements of the home market and military needs. To this was added the demand occasioned by the scarcity of butter in Britain, where for the first time in history margarine consumption outstripped that of butter — by 100,000 tons.²⁸

For the Borden government several other factors had also to be taken into account. There was growing agitation to have margarine available as a cheap and wholesome alternative to butter. However, fears were being expressed that the already limited supply of edible fats, such as lard, would be further constrained and that already high prices could be driven higher. The farmers were not inclined to increase dairy production because of a shortage of labour, and it was feared that competition from margarine would provide the farmer with more reason for curtailing milk production.²⁹

In response to the need for more food the federal Department of Agriculture adopted the position of the butter lobby, and, *vis-à-vis* margarine, they assumed a very cautious stance.³⁰ The catalyst on the issue proved to be the Food Board, which was generally inclined to see matters through the consumer's rather than the producer's eyes. On October 18, 1917, the Board's milk committee determined unanimously to recommend the legalization of margarine. The weakness of the argument, strategy and tactics of the butter lobby is evidenced by the fact that the one dairyman on the committee, Bingham of the Ottawa City Dairy, was in the area of the dairy industry which would be least directly affected by margarine's presence in the marketplace. To make matters worse, the more powerful voice of the lobby on the committee, W.A. Wilson, Dairy Commissioner for Saskatchewan, had, for unexplained reasons, found it impossible to arrive for the committee's sessions until the morning of the 19th, when he faced a *fait accompli*. A bitter Wilson returned to Regina.³¹

On October 23, 1917, the Borden government issued an order-in-council, P.C. 3044, making importation, manufacture and use of margarine legal, effective November 1, 1917. The product was to remain in the marketplace until such time as the government determined the "present abnormal conditions" to be at an end.[32] The lifting of the ban was thus to be temporary, with a pledge to that effect having been given to the dairy industry.[33] (This point was further emphasized when the 1919 act replacing the orders-in-council was entitled a "temporary" act.) P.C. 3044 required that paragraph (a) of section 5 of the 1914 Dairy Act — the ban — be suspended. Margarine could contain no foreign colouring matter, and no more than 16% water. Manufacture and importation were to take place only under license from the Food Controller. The Minister of Agriculture was to set up regulations governing manufacture in Canada. Strict records of the import activities were to be kept available for government inspection. Premises used for butter-making could not be used to make margarine, and butter makers could not be licensed to manufacture margarine. No preservatives other than salt were to be allowed. The Food Controller received the power to regulate the price of margarine. At the retail level, packaging and wrapping were required to be labelled "Oleomargarine" in letters not less than 3/4 inch square. Public eating places serving margarine had to display prominently a placard to that effect. No customs duties were to be levied on any imports complying with government regulations. Penalties for violations were set out. On all important points, these clauses met the proposal put forward by the Minister of Agriculture on October 17 — except in the colour and enforcement areas.[34]

On November 17, W.J. Hanna, the Food Controller, issued more specific regulations under which the importation and manufacture of margarine was to occur.[35] Four days later P.C. 3236 was issued,[36] solving the colour problem through tighter regulations. The original order-in-council stipulated only that margarine could contain "no foreign colouring matter," which opened the door for the American practice of including a capsule of colouring material to be mixed into the margarine by the consumer. The new regulation continued Hanna's order prohibiting the importation of colouring matter in margarine packages, as well as forbidding manufacturers, wholesalers, and retailers from dealing in or giving away or selling colouring substances. A shift of the enforcement of the regulations from the office of the Food Controller, seen as the consumer's ally, to the Department of Agriculture, appreciated as butter's ally, was also a victory for the butter lobby. On July 18, 1918, P.C. 1722 superseded the earlier orders,[37] being more specific as to how margarine would be dealt with, so that the purity of the product would be maintained and that fraud, feared by some to be inherent in the situation, would not be allowed. The tenor of all these regulatory acts expressed the butter lobby's perception that margarine manufacturers could not be trusted to provide a wholesome product; hence, the public must be protected.

No great fanfare accompanied the arrival of margarine in the marketplace. The *Moncton Daily Times* paid no attention; the Charlottetown *Guardian* gave brief notice but no editorial comment.[38] The *Digby Weekly Courier* recognized a long-range historical trend of heightening food crises; it was inevitable for people to seek out all available food stuffs. Margarine would find its permanent place in the food supply.[39] The response of the *Maritime Farmer* was illustrative of the best light to be cast on the situation by a supporter of butter: "It appears as if the substitute were to be almost 'regulated' out of existence before it starts—and this is as it should be.... If the regulations are lived up to, we will be disposed to congratulate whoever is responsible for making the best of a bad job—a very bad job."[40] Central Canadian newspapers paid some attention to the subject, but published no editorials.[41]

A sizeable group of Canadians had already been legally in contact with margarine. The ration scales of the Canadian armed forces included two ounces of butter.[42] However, these scales applied only to service personnel on duty in Canada. During the time that the 650,000 women and men served overseas, they were considered part of the Imperial armed forces and were on the British scale, which could include margarine. Information as to what foods the Canadian supplement consisted of is not available.[43] Given the human propensity to complain about institutional food, be it hospital, school, prison or armed forces, it is highly unlikely that the reports noting no complaints from service personnel[44] can be taken literally. British civilians eating the same margarine certainly did complain.[45] Although it was generally accepted that British margarines were of better quality than their North American counterparts, wartime British margarine could not have been an attractive food. Manufacturers had no choice as to the oils they could use: they were lucky to get any at all.[46] Margarine probably did not generate any more "bitching" by the soldiers than the butter supplied. One overseas veteran, sixty-five years later, could not recall whether he had eaten margarine, nor anything of his fellow soldiers' opinions of the food. All he could remember was that during fourteen months of his convalescence, the hospital in Britain had served margarine to the less seriously wounded patients, while he, gravely wounded by artillery fire, had been on a special diet which included butter.[47] After the war, the veterans' experiences with margarine were to become a factor in the debate whether or not to reimpose the ban.

Initial supplies of margarine took until the Christmas, 1917, season to arrive in Canadian retail stores.[48] Of the fifteen companies requesting licenses to manufacture, two, Harris Abbatoir and Swift-Canadian, both of Toronto, were initially granted approval.[49] From a marketing standpoint they were excellent choices. William Davies Company, which owned a controlling interest in Harris, and Swift were the two major packing houses in Canada and fierce competitors.[50] Both had national channels of distribution. Set-up costs and time were also minimized. Harris, having connections with Armour, the American packing company, since before the war, would manufacture Armour brands of margarine under license, presumably leasing Armour's equipment. On the other hand, Swift could simply transfer some of its margarine manufacturing operation from the U.S. to Canada. Thus, neither company had to commit vast amounts of capital or time to what was to be only a wartime industry. It is also of significance to note that these first margarine producers were companies which saw additional profits in the uses for their slaughter house by-products.

The 1917 version of the product was much better than the margarine which had been banned in 1886. Granted, some beef fat was still used in the manufacture of margarine, but a technological breakthrough in the early twentieth century now enabled manufacturers to use vegetable oils. The hydrogenation process, which revolutionized the margarine industry, involved the adding of hydrogen gas to heated oil while the oil was under pressure.[51] The oil hardened and as a result had a higher melting point and became spreadable. The first hydrogenation plant was established in the U.S. in 1910 and by 1920 hardened vegetable oil was the principal ingredient in margarine.[52]

Federal regulation came through the provisions of the Meat and Canned Foods Act, and the inspection system already in place under this Act. Margarine manufacturers were even more ready to step beyond the law than their butter counterparts. The ratio of convictions to poundage used was 9 1/2 times greater for margarine than that for butter.[53]

Two types of margarine, oleo and vegetable, were sold at this time. From 1917 to 1920 beef fat was still the main ingredient in Canadian margarine. Although the manufacturing process had been refined, packing-house waste was still used and the dairy lobby would exploit this fact to its fullest to get the ban reimposed. According to F.W. Wilder, a former superintendent at Swift, first grade margarine was made from the oil refined from the following fats: caul, ruffle, caul pieces of gut end, briskets trimmed from the bed pickings, crotch trimmings from the bed pickings, paunch trimmings, pluck trimmings, reed trimmings and heart casing.[54] This information was provided to the public after the war by the dairy industry in an effort to revive the 1886 health argument. Faced with a serious fat shortage during the war, however, the Canadian consumer welcomed margarine, and by September 1918 margarine too was in short supply and had to be unofficially rationed.[55]

After the war margarine manufacturers were better able to match the flavour and texture of butter by using vegetable oils. Coconut oil was popular because it was relatively inexpensive, it could be deodorized to a perfectly neutral flavour, and it melted quickly on the tongue.[56] Margarine manufacturers, however, used a variety of ingredients in an attempt to duplicate butter. For example, a popular American brand, Nuco, was made up of 76% coconut oil, 5% peanut oil, 2.5% milk solids, 2.5% salt and 14% moisture.[57] Margarine manufacturers had accepted the fact that butter was the gold standard and set out to create a product which would have its magical qualities. Hydrogenated vegetable oils went a long way toward making this possible.

In the 1920s, the margarine industry used some rudimentary marketing techniques in an attempt to garner a share of the Canadian market. Armour sold two margarines with two distinct market segments in mind. 3X Oleo-margarine was sold on a price basis with the main thrust of its promotion being to save money, while quality was a secondary concern.[58] This margarine, which was made from beef fat, was apparently sold to consumers who could afford nothing else and those such as boarding house or lumber camp operators who were price-conscious and could pass the product on to captive consumers. Armour sold Nut-ola, a vegetable oil margarine, on taste.[59] This margarine was intended for consumers who had a little more money but not enough for butter and those who preferred the flavour of margarine. When margarine was legalized, market segmentation would be used extensively to exploit butter's shortcomings and make inroads into the market.

Advertising was in its infancy in the 1920s. The food industry in particular was only beginning to realize its potential as a marketing tool.[60] Margarine had yet to win a legitimate place in the Canadian diet and as a result advertising was necessary to enlighten consumers to its potential advantages. Also, with two companies competing for the small Canadian market, some brand advertising was undertaken.

The advertising carried out by Armour indicates a basic understanding of market segmentation. As stated earlier, 3X Oleomargarine was sold on price and not quality; its slogan was "There's a Saving in Every Pound." Its copy also emphasized this theme. One advertisement stated: "Cut the size of your grocery bill by using 3X Oleomargarine. One pound will prove its wholesome goodness and its economy!"[61] Another maintained: "Particular housewives serve it on their tables and choose it for their favourite recipes. Because of the reasonable price it can be used freely."[62] Nut-ola, Armour's premium brand, stressed quality in its advertisements which bore the slogan: "A Table Treat." Its copy pointed out that it was made from "highly refined cocoanut [sic] and peanut oils, churned in pasteurized milk and salted to taste." Because of these ingredients Nut-ola had:

> *Flavor Unsurpassed* Makes a delicious and tasty spread of delightful flavor for bread, hot biscuits, griddle cakes, waffles, etc.
> *Quality Uniform* Churning, blending, working and salting of Nut-ola nut margarine are according to fixed formulas. The quality is guaranteed.[63]

Conspicuous by their absence in the period's advertisements are any references to dairying or comparisons to butter. This is presumably because all aspects of the margarine industry were still under strict government control, and no advertisements that blasphemed the good name of butter by comparing it to margarine would be permitted. In the U.S. and in Europe such was not the case, and slogans such as "churned especially for lovers of good butter," "made in the milky way" and "smart people can't tell the difference" were used.[64] In Canada, however, the margarine industry was forced to sell its product on its own merits.

The prime quality which put margarine on the dinner tables of Canada was its low cost. The price of margarine during and immediately following the war was subject to government approval, which accounts for the stability of its price over these years. Although margarine prices did not rise with butter prices during the war, after 1921 they followed them down.[65] Margarine could be sold only if there was a substantial saving over butter. This sales threshold was apparently 50% of butter prices; if margarine prices rose above this level consumers chose butter instead. The initial price gap was much smaller. On December 31, 1917 butter sold in Kitchener, Ontario at the retail level of 47¢ per pound and margarine at 35¢ per pound.[66] This phenomenon was evident in other parts of the world where the threshold ranged from 35% to 55%.[67]

The total consumption of margarine was minuscule (3-3/4% or, on a per capita basis, 5.9 pounds of margarine to 156.2 pounds of butter) when compared to the amounts of butter which Canadians used in the same period (see Appendix 7). While dairy butter production suffered a decline of 25% over the whole decade, that of creamery butter almost doubled. Margarine production and importation made up most of the shortage in domestic supply created by the extra effort made to meet British wartime needs for butter. Decline in dairy butter production was not attributed to the competition from margarine but rather to the difficulty of procuring farm help and the wider use of creameries. The consistency of the consumption patterns — for margarine a steady decline, for butter a continuous increase — as well as the wholesale price contrast led to the conclusion that margarine did not affect butter sales. Canadians appreciated margarine only as a contributor to the resolution of an emergency, not as an acceptable substitute. In 1917, margarine could aid the efforts to expand the Canadian war effort by enabling Canada once again becoming a net exporter, rather than an importer, of butter (see Appendix 6).

The steady growth of creamery butter production also demonstrates that its fortunes were not adversely affected by margarine's competition. Production continued to expand beyond 1924, the year margarine was removed from the market.[68] The conclusion reached by the Alberta Dairy Commissioner for the first and heaviest year (1918-19) of margarine's competition is interesting. Not one reference to margarine is made in the report. The Alberta creamery butter producer enjoyed a "rising" market and "a comparatively high price elevation" was reached. The average price of 45.3¢ per pound for the May-September 1918 season was 15% higher than that received the previous year, and 75% higher than in 1914.[69] The Saskatchewan Dairy Commissioner also reported average prices higher than ever before in world history.[70] Butter did not suffer from the competition with margarine.

Taste was really a secondary factor. A contemporary has left his impressions: "In texture it was very firm, but rather tallowy, and although the initial impression was not unlike the genuine article [butter] a sensation of something different with a flavour of staleness, approaching rancidity, lingered on the palate."[71] But by comparison it was as good or better than all but the top-grade butter.[72]

Economically, margarine achieved nothing more than taking up a small corner of the market. However, the butter industry reacted as if the threat were much more serious. Some cases of panic selling of butter by dealers developed, once the orders-in-council were published.[73] Under W.A. Wilson's leadership the lobby pressured federal authorities to issue more precise guidelines as to who would supervise the new regulations. As a result, the federal Commissioner of Dairy and Cold Storage was given the task.[74] A supporter of the butter lobby, he would make sure the margarine people played the game fairly. The extensive nature of the effort to protect butter is shown in the attempt to have the margarine display removed from the dairy section at the Canadian National Exhibition. From the perspective of the butter people, margarine ought to have been transferred to the manufactured food section alongside lard and sausage.[75]

Such efforts were still grounded in the myth of milk, in which most Canadians still believed. To counter such contentions, the margarine producers, in their substantial advertising campaign, concentrated on the three factors of similarity to butter in taste and colour, purity of product, and competitive price.[76]

The only debate in the House of Commons on the initial action to legalize margarine also demonstrates the sensitivity of the butter lobby. The geographical bias appeared again in that each of the six MPs speaking on the motion was from Ontario, all but one from ridings in the southern part of the province in which dairying was a strong factor. Francis H. Keefer, from northern Ontario, argued the strongest pro-margarine position by stating that regulations were already much more stringent than in the United States. Edward N. Nesbitt, from North Oxford, was the only member from the rural south to attempt to strike a balance. The other four argued the party line without presenting any evidence to the effect that the margarine interests were not playing by the rules. After the debate had released some of the emotional pressure, and Agriculture Minister Thomas Crerar had restated government policy, the motion was withdrawn.[77]

For the dairy industry, one positive result did come out of this crisis. It became obvious to men like Saskatchewan Dairy Commissioner, W.A. Wilson, that at least one factor leading to the failure of the butter interests to block the approval of margarine had been the industry's lack of proper organization to carry on effective lobbying.[78] Failure in 1917 and the felt need to regroup to reimpose the ban led to efforts to form a stronger national dairy organization and a national agriculture body. This effort is typical of the moves made by many interest groups in the direction of centralization during succeeding decades. Eventually, the Canadian Federation of Agriculture came into existence in 1936.

Margarine's initial presence in the Canadian marketplace came about as a patriotic measure. It did not threaten the dominant position of butter in that marketplace. The dairy industry's negative reaction was based on fear — a fear that would undergird its struggle to have the margarine ban reimposed. The legal presence of margarine, because it was a proven wholesome food, did force a re-evaluation of the basis for opposition. One of the bastions of nineteenth-century Canada's received culture, the rural life, needed to be protected.

CHAPTER FOUR

Renewed Exclusion 1919-24

Although the 1921 census showed a 50/50 ratio of urban and rural dwellers, the received culture of Canada was still agrarian in its orientation. Thus butter continued to bask in high esteem, a position not threatened by margarine's presence in the marketplace. Margarine's brief period of legality did serve to demonstrate its wholesomeness. But it was to be banned again in 1924, because of its potential threat to butter.

The butter industry came to the end of the war sharing the general Canadian optimism which was based on the existence of a world-wide food shortage.[1] It was understood that the European dairy industry was in chaos. The impact of the war had been devastating, and the millions of cows that had survived the war were now being slaughtered for food. The insatiable British butter market was seen as being over 200 million pounds short of matching demand. Canadian dairymen considered they had a great opportunity to take over that market from continental dairymen.

A second factor upon which the dairymen based their high hopes was the domestic scene. During the Laurier boom, Canada had experienced sustained growth in population and a concomitant expansion in the domestic market for milk. The variety of dairy products was becoming more complex: milk powder, condensed milk and ice cream were the leaders, taking a sizeable amount of fluid milk, and providing serious competition for the continuing production of the traditional products, especially cheese.[2] On the supply side, dairying, east of the Great Lakes, had reached the limits of its growth; in the prairie provinces it was just beginning to realize its potential.[3]

At the end of World War I, in spite of serious problems within the industry, optimism reigned. Inflation had played havoc with profit margins. The price of cows had risen 50% since 1914, the price of feed and labour, when available, 75%. The *Canadian Annual Review* vouched for the detrimental impact of inflation on the dairyman: between April 1914 and March 1919 the prices of nine dairy commodities had risen 78% while the cost of 22 commodities which the dairy farmer had to buy had increased 170%![4] Even a 30% increase in prices was inadequate, but the protests of consumers were so vigorous that further increases could not be contemplated.[5]

Within the perception of the dairy industry the only solution at hand was to remove margarine from grocers' shelves. Dairymen argued that the European dairy industry was mature, in that it had reached capacity and that large amounts of dairy products were being imported. On the other hand, the Canadian dairy industry was still in its infancy, and required protection from margarine until the western potential had been fully developed and maturity had been reached.[6]

In the fight against margarine, the nutritional arguments, of course, were brought to the fore and found a receptive audience still captive to the myth of milk. In the long run, it did not matter that margarine was not nutritionally inferior to butter, or that wartime consumption could not be shown to have been deleterious to public health.

The five years from war's end to butter's victory were years of frenzied activity for the leaders of the butter lobby. Percy Reed, W.A. Wilson's successor as Saskatchewan Dairy Commissioner, continued to make his office the focal point for Western Canada. E.T. Love, Manager of Woodland Dairy in Edmonton and Secretary of the Alberta Dairymen's Association, was employed as propagandist. The battle plan was basically twofold: get the correct information before the proper people, and provide evidence that the public supported the butter position. Resolutions from the National Dairy Council and from provincial counterparts were sent to Ottawa. Support was successfully solicited for resolutions from Manitoba creamerymen; from the Regina Council of Women, which reversed its stand of 1918 when the price of butter had been much higher; from the Homemakers Clubs; and from the Federated Women's Institutes of Canada. While the primary intent of this activity was to influence the national government, efforts were also made to generate action on the provincial level. Each of two such efforts failed: in Alberta by a 32-12 margin.[7] In Nova Scotia the effort got as far as support by the Assembly's agricultural committee; the Assembly itself never voted on the resolution.[8]

The message was to be spread through the established media, such as the old reliable *Farmer's Advocate*. To these were added such publications as a news-sheet entitled *Anti-Oleo*, published over the winter 1923-24, an obvious reply to a similar sheet put out by the pro-margarine interests.[9] The first issue of *Anti-Oleo* argued that the recipe for margarine was "G.O.K." (God Only Knows). Whereas the butter industry could declare the purity of its raw material, the margarine manufacturers were described as secretive about the contents of margarine, suggesting impure ingredients.

E.T. Love's pamphlet, *Oleomargarine And Its Relation To Canadian Economics: Shall We Foster A United States Industry Or Shall We Foster One Of Our Own National Industries—The Dairy Industry?* offered another example of the ammunition used in the propaganda war. Love developed the nutrition argument, incorporating the myth of milk, but showed a superficial appreciation of the role of vitamins. The prime thrust, in any case, was economic. The butter industry was built upon a raw material produced and processed totally in Canada, giving employment to thousands. Margarine was manufactured by two companies, one an American subsidiary, from imported raw materials. The margarine manufacturers had the great advantage of exemption of duty, while these same products, if imported to make shortening, paid duty. Thus margarine provided unfair competition not only to the butter maker, but also to the shortening manufacturer. Between 1917 and 1923, butter exports exceeded imports by nearly fifty million pounds. Over the same period forty-six million pounds of margarine had been imported or manufactured in Canada. The thrust of Love's argument was that margarine forced butter makers to find a foreign market for over forty million pounds of butter that Canadians would otherwise have consumed. It is worth noting that this sort of thinking was radically different from that found in other countries. For instance, Danes encouraged domestic use of margarine in order to expand butter exports. Love's pamphlet went on to argue that consumers were not being helped by margarine, at least not to the extent popularly believed. He noted that over 50% of the margarine consumed in Canada was used as a shortening instead of as a substitute for butter as a bread spread. Much of the rest was being served by boardinghouse keepers. Most of the poorer classes of Canadians were not the benefactors, as margarine's advocates had predicted.

> The exclusion of oleomargarine would work no hardship on the consuming public. A dollar's worth of butter is of more actual food value to the families of even the poorest working man than a dollar's worth of oleomargarine, and it is of vital importance to the working man of this country that his dollars be spent on all-Canadian products rather than for goods imported from outside our country already manufactured, or for goods manufactured in our country from products which were produced outside of it.[10]

The protectionist argument could hardly have been put more strongly. Love continued by formulating arguments specifically aimed at the national political parties. He recognized that the Liberals, under Mackenzie King, were in power in Ottawa, somewhat on the sufferance of the Progressives. To the Liberals, he pitched the argument that their party stood for the lowest tariffs consistent with Canadian interests; but in reality this could also mean practical protection. A national industry needed protection from margarine's competition. The Progressives were given the argument that their constituents would also be served by the continuing existence of a strong dairy industry as part of a national mixed farming programme. The Conservatives, in opposition in the Commons when Love wrote, were ignored.

Oleomargarine And The Physical Deterioration Of The Working People: A Plea For The Health Of The Child Of The Nation,[11] written by Alfred Farmilo, followed a similar theme. Farmilo was an organizer for the American Federation of Labour in Western Canada and President of the Edmonton Trades and Labour Council. One would expect that as a leader of the working class he would be arguing in favour of margarine. Since he was not, it might be worth noting that Farmilo's position was the same as that of another labour leader, William Irvine, Labour MP for Calgary East. Irvine gave as the reason for his objections that margarine was a second-class food. Labour ought to be seeking adequate wage scales to enable workers to purchase quality food.[12]

Farmilo's line of thought followed that of Love, but it was steeped more markedly in the myth of milk. Dedicated to "the children of Canada," it closed with the plea to "raise our voice in a national protest against those who would, at this time, subvert the future health of our nation for a present monetary gain, through the substitution of oleomargarine for pure, wholesome butter in our Canadian households!"

Such material was sent to at least 2,500 people across the country, including all members of the House of Commons and the Senate, 800 creamerymen, 126 members of the National Council of Women and the United Farmers of Alberta executive directorate, the Canadian Council of Agriculture, all of the press of Alberta, and the daily press across Canada.[13]

The supporters of margarine were equally active in making their case. Their campaign made use of essentially the same tactics as those of the butter supporters. Petitions were put before parliament.[14] Consumer groups, such as the local Councils of Women, agitated in support of margarine.[15] The Canadian Pacific Railroad published a statement of support in its journal.[16] Trade journals such as the *Canadian Grocer* were widely used as vehicles to spread the message. An extensive magazine and newspaper advertising campaign was carried on.[17] Armour and Company promoted its two brands of margarine: 3X, primarily because of the savings to be gained[18] and Nut-ola, because of its flavour and quality.[19] Non-food trade journals were used to promote the argument in favour of extending margarine's life.[20] Swift-Canadian, the other firm manufacturing margarine, lobbied by mail with trades and labour councils, women's clubs and the Great War Veterans' Association.[21] Company spokesmen were also in direct contact with members of parliament.[22] Resolutions were sent to the federal government from the Ottawa Board of Control, the Ottawa Imperial Order of the Daughters of the Empire (IODE) and the Canadian Women's Christian Temperance Union.[23] Literature issued included a pamphlet entitled "why ban a good food?"[24] and a news-sheet entitled "Oleomargarine" issued over the 1923-24 winter.[25]

The industry's argument centred on the issue of wholesomeness or, in other words, protection of consumers. Seventy-five percent of the ingredients were the products of Canadian farms: oleo oil, neutral lard, milk and butter, which had to meet strict government inspection standards. The remaining 25% were imported vegetable oils which also had to meet government standards, and salt. A bow in the direction of the myth of milk, in its more scientific form, was also made:

> "Growth accessories" is the name given to three substances which have, in recent years, been found to be highly important parts of our daily foods.

Their functions are known to be vital to growth, but they have not yet been fully studied.... Not only are the growth elements, including vitamins, present in the oil, but they are in milk with which oleomargarine, in the process of manufacture, is churned.[26]

Although both sides lobbied, the greater energy was expended by the butter lobby. It wanted to make sure it would not lose the battle for lack of effort, as W.A. Wilson judged to have been the case in 1917. This perception proved correct, as the sheer volume of pro-butter propaganda falling on ears attuned to the myth of milk achieved the desired result.

The margarine lobby made a major, though not an all-out, attempt to reverse the verdict because, in spite of the logic in its arguments, margarine had not caught on with the public. About half was being used by the baking industry, where a greater consistency in the end product was appreciated. Much of the remainder was consumed in boarding houses, lumber camps and on fishing boats as a cost-cutting device in a captive consumer situation. Some middle-class consumers used it in the home. As to public eating establishments, no objections were raised when an MP commented that he had yet to see in any such place the required sign signifying that margarine was being served.[27] The poor, for whose relief it had been argued margarine should be legalized, were not responding to the extent expected.[28]

Thus, propaganda to educate the parliamentarians came from two sides. The issue was to be a very sharp thorn in their sides for several years. The question of protection of Canadian industries — generally by means of a protective tariff — divided Canadians sharply in the inter-war period. The Conservatives, under Robert Borden and Arthur Meighen, stated the rationale in favour of high levels of protection. The position of the Progressives, the new political force on the national scene, illustrated even more the complexity of the tariff and margarine issues. They gave voice to the farmer's strong laissez-faire belief when it came to manufacturing. But as for applying that belief to their own industry — dairying — many were prepared to exchange consistency of principle for protection of pocketbook. The Liberals, under Mackenzie King, preferred a more pragmatic approach, but recognized that the country was split and attempted to downplay the issue whenever possible. It was during this struggle over the fate of margarine that Mackenzie King was to acquire his irritability on the subject. His sensitivity was to be a major factor in confounding attempts to find a permanent solution after World War II.

In response to inflation, which continued to cause serious problems after the end of the war, the Meighen government adopted the tried-and-true tactic of appointing a select committee of the Commons to inquire into the matter. It is worth noting that one of the committee members was W.D. Euler, the MP for Waterloo North. Although his role at this time demonstrated no specific interest in margarine, it could well have kindled Euler's concern for the effects of inflation on people's lives, setting the stage for his fight to help Canadians confront threats to their standard of living in the late 1940s. The general conclusion of the committee was that there had been no undue profit-taking in the food industry. War conditions in general had been the vital factor in price inflation. In the committee's report, no specific mention was made of margarine or its role in the marketplace, but it was the focus of some questioning. Representatives of both the creamery and the agriculture interests agreed that, for the general good, margarine should remain available to Canadian consumers so long as the price of butter remained at a high level. The only need was to protect the consumer from fraud. However, agricultural spokesmen argued that when more normal times and prices returned, butter should not be required to face margarine's competition.[29]

The result of the committee's work led, in the autumn session of parliament, to the Meighen government replacing the orders-in-council, which had given margarine a temporary wartime status, with an act extending the legal limits of manufacture and importation until August 31, 1920 and of sale until March 31, 1921.[30] The ensuing debate was again dominated, by an 8 to 6 margin, by Ontario MPs, of whom all but two were from southern, rural ridings. The contextual framework had changed, since margarine was now legally in the marketplace. The thrust of the debate remained basically the same: protection of the dairy industry and of the consumer. There was some complaining about the inequities in the regulations. For instance, a creameryman was forbidden to make or even have margarine on his premises, but nothing controlled the margarine manufacturer when he worked butter into his product to give it flavour and colour close to those of butter. The best grades of margarine contained 15% butter. Manufacturers admitted buying all the June butter they could acquire because it was the most highly coloured. Donald Sutherland (Conservative, South Oxford, Ontario) called for a reimposition of the ban so that the Canadian dairy industry could regain its strength as one element of a country now returned to peace. But, in his next breath, he admitted he would be satisfied if the law were changed to prevent any incorporation of milk or butter into margarine. Wilks Baldwin (Liberal, Stanstead, Quebec) pointed out that the American experience had seen the competition from margarine lead to a sustained increase, rather than a decrease in butter prices. Thomas Crerar (Progressive, Marquette, Manitoba) suggested that if the butter industry wanted to continue in its efforts to compete overseas, it would also have to meet margarine's competition in that broader marketplace. Therefore, there was no reason why butter and margarine should not compete in Canada. With butter wholesaling in Toronto at 53¢, and margarine at 37¢ per pound, high prices by anyone's estimation, the Baldwin-Crerar arguments won.[31] Under these conditions margarine entered the peacetime market.

The parliamentary debates on margarine over the five years to 1924 were really one continuous argument—with little new of substance and much repetition. Without a doubt the major theme through the arguments of the parliamentarians was protection, but as each side perceived it within their cultural context. A permanent position in the marketplace for margarine was not in the national interest as defined by the dairy industry: it enabled the consumer to be deceived. It would be detrimental to the fortunes of one of the mainstays of Canadian economic and social life, and it profited only "big business," one of the enemies of that mainstay. Margarine's supporters also argued protection of the national interest: a wholesome food at a cheaper cost. In 1921, W.D. Euler, joining the debate for the first time, delivered one of the best speeches on the subject. He set out what was to be the substance of his case over the years. The only reasons for dispensing with any product in the marketplace was that it was unhealthful or otherwise unfit for human use, or was being sold in a manner through which the public was liable to be deceived. Margarine was not guilty of either of these charges. Any individual had the right to purchase those products which could be afforded. Euler added the less fundamental point that the present economic circumstances were such that many Canadians could afford only the cheaper margarine.[32]

William C. Good, dairy farmer from Brant, Ontario and a rising star in the farm movement, had been elected as a Progressive to the House of Commons in 1921. Good made one of his first efforts to influence the affairs of parliament in his support of the pro-margarine position. He objected to the protectionist principle underlying the butter forces' stance.[33]

The debate took place in the context of the very fluid political scene of postwar Canada. British Columbia's Dr. Simon F. Tolmie, Minister of Agriculture in Borden's and Meighen's Unionist ministries from August 1919 on, was a veterinary surgeon and a dairyman. He argued consistently, as private member and as minister, that margarine was no threat. So long as margarine's ingredients were wholesome and there was no consumer deception by misrepresentation or false labelling, the cheaper product had its place.[34] Most likely he would have supported permanent legalization of margarine but, appreciating the political realities, he acted only to extend margarine's life on a year-to-year basis.

The 1921 election was significant in that it brought out the regional nature of each party's support, as well as creating a situation in which, for the first time, no party could claim a parliamentary majority. The national significance of the margarine issue is further illustrated by the fact that the 1921 segment of the margarine debate was the first in which Ontario MPs did not dominate.[35] The Liberal victory brought Tolmie's replacement by William R. Motherwell, a Saskatchewan farmer who was firmly on the side of butter. Motherwell was politically strong enough to overpower Mackenzie King's fence-sitting and Finance Minister W.S. Fielding's active support of margarine.[36]

The Commons expressed an unequivocal position when in 1922 it rejected A.W. Neill's (Independent, Comox-Alberni, B.C.) resolution (57-83) calling for the reimposition of the margarine ban. Yet the government declined, under Motherwell's pressure, to recognize the significance of this action. The weapon he found to swing the Prime Minister to his side was the pledge by the Borden government that the suspension of the ban was temporary, due to war exigencies. This was his strongest argument, for it enabled King to act so as to blame his opponents, and take to himself what credit was due. Motherwell struck another sympathetic chord with King with the argument that honouring the pledge would foster national unity. Eastern Canadian dairy interests had had thirty-one years of protection to build up their industry. Western Canadian dairy interests now wanted an equal chance to get established.

In the broader context Motherwell's most significant action was to have parliament permit the importation, manufacture and sale of renovated butter. Most Canadian dairy butter was being exported to the United States for renovation and sale. The government could thereby point out that a suitable, cheaper substitute for creamery butter would be in the marketplace to take the place of margarine.[37] Here was exactly the sort of innovative approach the butter industry should have been seeking and supporting in its efforts to meet margarine's competition. Neither Motherwell nor the industry seem to have been concerned about the heavy volume of fluid milk and cream being exported to the United States. Retaining it for Canadian creameries to turn into butter would have increased annual production of butter about 15%.[38]

Despite the strength of Motherwell's arguments, King must have winced when he met Arthur Meighen, leader of the opposition, in a classic King-Meighen duel. To Meighen, a temporary pledge certainly did not bind future governments to any particular action. A high tariff should be imposed so as to prevent the importation of margarine and thus protect the butter industry. Such action, plus stricter regulations to avoid consumer deception, was all the dairy industry required.[39]

Motherwell's political concern for Western Canada was well founded, for on the prairies the Progressives had taken 40 of 43 seats in 1921. The 1921 election brought the Progressives to Ottawa in numbers sufficient to let them be the official opposition had they wished it. The margarine issue placed these MPs on the horns of a real dilemma. On the national level this paradox showed itself in the fact that while the dairy farmers were doing all in their power to destroy margarine, the grain growers could oppose such action.[40] In 1920, the application by the National Dairymen's Association for membership in the Canadian Council of Agriculture was rejected. The reason given was that entry would split the interests of the CCA. The present members, as grain growers, appreciated free trade as they sought to sell their grain in the world's markets. They sought government protection of their livelihood by means of freight rate subsidies such as those embodied in the Crow's Nest Pass Agreement. The dairy men, with an internal market, favoured government action in the form of a protective tariff or outright prohibition.[41] On the political level the prime problem for the Progressives was how to serve a constituency that, on the surface, was homogeneous, yet fundamentally varied.

Mackenzie King was to recall these years as times in which his party had "accomplished much despite hardships of a most exceptional kind"[42] and Euler was to remember the debates as warm and bitter.[43] However, much clearer perceptions of the occasion were held by the Ottawa correspondent of the *Winnipeg Free Press* when he labelled the topic "hackneyed."[44] The *Canadian Annual Review* also brought some ironic perspective when it pointed out that the burning issues of the parliamentary session were over two private members' bills to abolish betting at race tracks and to legalize the sale of margarine.[45]

The new King government dealt, in the first instance in 1922, with the margarine issue simply by continuing the practice of extending its life by one year. On this occasion the Progressives split less drastically than did the other two parties. Thirty-five followed the dictates of the principle of free trade and helped defeat A.W. Neill's motion to discontinue manufacture and importation of margarine after September 1, 1922. Six Progressives voted profit over principle. The Conservatives split 16-7 in favour of reimposing the ban. The governing Liberals were also unable to avoid a major rift in their ranks, one of the first which King's government had to contend with in its minority position: King himself was in Europe. The Liberals voted 42-24 in favour of the ban.[46] From another point of view, that of urban-rural, the vote in the Commons showed the increasing strength of the urban sector. The MPs from predominantly urban ridings divided 20-5 in favour of extending margarine's life. The MPs from ridings with significant urban and rural elements[47] voted 18-8 in favour of extension. The 45-44 split in favour of extension among the rural members demonstrated the mixed opinions of these people.

The following year, the Progressives faced the economic and political realities of their constituents with less idealistic fervour. Thirty-eight became practising protectionists. The Rev. T.W. Bird (Nelson, Manitoba), one of the purists, expressed his sympathy for his fallen comrades:

> I have a great deal of sympathy with some hon. members on my own side of the House who imagine that by prohibiting oleomargarine they are doing a service to the cause of agriculture. However ... it seems to me that the farmers of this country are beginning to nose around the protection trough, ... are beginning to see the attractions of that trough and some of their old time prejudices are being held less firmly than before. Will any man stand up in this House and say that they are to be blamed for that, that they are swallowing their principles? Why, what are principles for but to be swallowed ... ?
>
> It is not the swallowing that is wrong; it is the inopportune time that we choose to swallow.... I want to point out this, and I hope I do not need to point it out because everybody with his eyes open will have seen it before now, ... we have seen the British farmer go to his government with the pleas for protection; we have seen the United States farmer go to his government with the plea for protection, and we have seen the Canadian farmer tonight not exactly asking the government for protection, but just hovering around and trying to hint to the government that they would like some protection if they dared give it to them. That is the position. Now where are we coming to? ... we are just beginning to get in the thin end of the wedge and by and by you are going to have a Farmers party in Canada that will be one hundred per cent protectionist. They will all be in the trough then.[48]

While the Conservatives mustered a bare majority (18-16) in favour of renewing the ban, the Liberals were able to keep the break in their ranks to a small but still significant number (18-63). When viewed from the urban/rural delineation a truer significance shows. The rural members divided 97-22 in favour of rejecting margarine, while the urban/rural representatives also rejected it, by a 21-13 margin. Even the urban vote swung 19-6 against margarine. Provincially only Alberta (5-6), the Yukon (0-1) and Prince Edward Island (1-1) did not support the reimposition of the ban. The margarine issue was a vital one for the Progressives as it marked the turn in their history from "uncompromising foes of privilege in all forms" to "sensitive and faithful representatives" of agrarian and sectional interests.[49]

Margarine lost to butter for no rational reasons but because of fear. The dairymen were concerned about future competition. The mind-set of the farmers and the community at large was still rurally oriented, so the politicians felt that they had to respond to protect the interests of society as a whole. Margarine would remain banned in Canada until the demographic pattern was such as to give the urban element a majority. The next generation, urban in its orientation, would act to apply its standards of values to society.

CHAPTER FIVE

Holding the Line 1924-45

The brief relaxation of the ban on margarine had come about because of wartime exigencies. During this short period of contact some Canadians had had the opportunity to perceive it as a healthful food, but the impact had been minimal. By 1924, Canadians were prepared to accede to the apprehensions of the dairy industry and to reimpose the ban. In the longer perspective, however, the circumstances of economic depression, war and increased urbanization would generate the conditions in which the ban would be removed permanently.

During the inter-war decades, Canada's population increased by nearly five million: much of this increase was found in the urban sector. The Depression broke the steady growth of urbanization, but by the end of World War II, six of ten Canadians were living in an urban environment. The demands of two wars also intensified the extent and the complexity of industrial growth, with the concomitant pattern of the urbanization of Canadian society.[1]

Urban growth was a significant factor in the history of margarine. It meant a larger market for dairy products, as well as a greater demand for a wider range of products. It also meant the appearance of a generation further from its rural roots than its predecessors, and therefore its received culture did not necessarily include a strong allegiance to the myth of milk. The groundwork was being laid for the acceptance of margarine in the years following World War II.

If the number of milk cows in Canada is used as a standard, the dairy industry expanded almost continuously from World War I until 1934, when the size of the national herd reached its all-time peak, and remained at that level for four years. From 1937 on, there was an almost continuous decline in herd size until 1952,[2] while ongoing efforts to improve per-cow output ensured that milk production continued on an upward trend until after 1960.[3] Production of dairy butter steadily declined, while that of creamery butter followed an opposite pattern — the combined total of butter produced followed a gradual downward trend throughout the 1920s and the Depression, and only World War II brought about a reverse in the pattern. More butter than cheese was produced consistently but these foods were joined in the marketplace by a broad range of milk-based products: evaporated and condensed milk and skim milk powder in 1917 and ice cream in 1920.[4]

Per capita consumption statistics showed a fairly regularly expanding use of milk in its fluid form as well as of its by-products. Even the hard times of the Depression failed to have a significant impact: butter consumption fell slightly during the first two years of the Depression before resuming the growth pattern evident since World War I.[5] The steady rate of consumption was helped by the fact that fluctuations in the wholesale price of first-grade butter paralleled the ups and downs of the consumer price index.[6] The average person seems to have been able to continue eating butter; it was the poor who had to forego any bread spread or use substitutes, such as lard.[7]

From 1916 to 1948 the Dairy Branch of the Saskatchewan Department of Agriculture maintained an extensive file on the margarine issue. When its files are taken as a guide, 1926 saw the last efforts by either side, outside of the public sector, to influence the situation before World War II. At the instigation of Mrs. Elizabeth Shortt of Ottawa, the National Council of Women in Canada was urged, at its June 1926 national meeting, to pass a resolution seeking the renewed legalization of margarine. The Saskatchewan, Alberta and Manitoba delegates were able to block the pro-margarine forces at the committee level. It was argued that although the economic interests of the poor would be served if margarine were sold, butter was nutritionally superior and its consumption would foster the health of the consumer, especially that of children — a demonstration of the generational gap in knowledge of the relative nutritional value of the two foods. The change of ministry in Ottawa, after Arthur Meighen's victory in 1926, raised some question whether the new government would be more ready than King's ministry to resurrect the issue, a concern that was sidetracked during the constitutional crisis that led to Meighen's defeat four months later.[8]

The House of Commons continued as the focus of some manoeuvrings regarding butter's fortunes in peacetime. In 1925, W.R. Motherwell, the Minister of Agriculture, saw through the House a bill amending the 1914 Dairy Industry Act by, among other housekeeping elements, bringing into legal existence a definition of margarine so that the fraud element could be more effectively policed.[9] The efforts of the National Dairy Council to gain stronger protection of the Canadian market from a growing flood of cheaper Australian and New Zealand butter were stonewalled by the King government until 1930. In preparation for the election later that year, Charles Dunning, Minister of Finance, increased the tariff on butter to 4 cents for the British preference, 6 cents for the intermediate tariff, and 7 cents for the general rate. An anti-dumping clause was also added.[10] Such action was not enough to help avert the disaster for the Liberals at the polls on July 28.

The Conservative ministry of Richard Bennett responded to campaign promises by doubling the tariff rates, arguing that Canada's climate imposed additional costs of production in contrast to the industry in more temperate climates, and this disadvantage needed to be offset. Secondly, this "oldest" of Canadian farm enterprises, with all of the jobs related to it, was worth saving.[11] Bennett's government also began a series of direct subsidies to producers and manufacturers in order to avoid a rise in prices that would force poorer Canadians to stop buying butter. Revoked by King's government when back in office, the subsidies were to be restored in wartime.[12]

Thus during the Depression, the government and the butter industry gave up attempting to produce a surplus for export and moved to a policy of self-sufficiency and of protection of the home market for Canadian producers. Within ten years the question, whether even the domestic needs could be met, would be raised.

The nutritional improvements in margarine made in the inter-war years had no effect, in the Canadian marketplace, on its competition with butter. But then, even in Britain, the old prejudices against margarine's less nutritious character did not disappear until after World War II began. In fact, by 1935 the quality of manufactured vitamin concentrate was of a consistency that enabled Unilever to add the concentrate to all its lines. Now margarine had the advantage, for its vitamin content could be held constant, while in butter the natural content of vitamins A and D fluctuated with the seasons.[13]

Although direct competition between butter and margarine was not to be, the butter industry maintained a watch against any sort of attack from whatever source. In 1934, for instance, an amendment to the Dairy Industry Act extended the prohibition of the importation, manufacture and sale of butter substitutes to those containing mineral fats. This action was taken to protect the Canadian dairy farmer and the consumers from a new butter substitute which had already appeared on the American market.[14] No evidence has been discovered to establish whether any fraud related to the mineral-based butter substitute was uncovered in this era.

The dairy industry fought hard to keep substitutes out. But it also continued to strive energetically to maintain credible standards across Canada for its own product. For example, in 1937 the Prince Edward Island legislature, on petition from the provincial dairy association, enacted the dairy product grading regulations under which federal inspectors graded all butter in interprovincial trade. This action brought the province into line with most other Canadian provinces.[15]

The industry spent several years attempting to get tariffs put on vegetable oils imported for the manufacture of shortening, in the hope of increasing the domestic market for the production of butterfat. These efforts did not win the support of all those involved in dairying. For instance, P. Reed, secretary of the Saskatchewan Dairy Association, opposed them as most likely to generate demands for the legalization of margarine. Reed preferred to "let sleeping dogs lie."[16]

When war began in 1939, Canadians were again faced with a two-fold food problem: to feed themselves so that they could perform their wartime tasks, and to expand food production so as to meet the needs of the armed forces and of their British allies. This task was to be accomplished within the framework of a much more sophisticated appreciation of nutritional requirements.[17] The continuing high reputation of milk, the "food of foods,"[18] was based on solid nutritional arguments; as well, the myth of milk still carried some strength in the minds of Canadians.

Butter and cheese were two of the foodstuffs Canadian farmers concentrated on producing. Greater efficiency led to expanded milk production despite dairy herds diminished in size from their 1934 peak.[19] Milk production increased from 15.4 billion pounds in 1940 (the low point) to a peak of 16.8 billion pounds in 1944. Butter production followed the same pattern: 6.2 billion pounds in 1940, 7.3 billion pounds in 1945. Cheese production kept pace with a low of 1.61 billion pounds in 1940 and a wartime high of 2.29 billion pounds in 1942.[20] Sizeable portions of this cheese production went to meet the commitments made to Britain. Contrasted to a Depression low of 55.7 million pounds and a high of 90.9 million pounds, cheese exports in 1942 were half again that of the Depression high point. The 1943 peak was 700% higher than the quantity sent to Britain in the first year of war.[21] This extra effort in 1943 was a response to the loss to enemy submarines of two ships carrying Australian and New Zealand butter.[22]

The absolute increase in income was the cause of increased use of milk in all forms among civilians as well as military personnel.[23] The all-time per capita high of 1,244 pounds was reached in 1942. The per capita consumption of cheese steadily expanded beyond this period. Butter use reached its peak in 1942; fluid milk use peaked in 1945.[24] Production of cheese was still more profitable for the farmer. From 1943-44 to the end of 1947, the federal government paid countervailing subsidies on various milk products: butter production received the largest amount, a total of $84.1 million; fluid milk $44.5 million; cheese $14.2 million, and milk (for concentration purposes) $5.9 million.[25] The purpose of this intervention was to offset the fact that Canadians had one of the highest butter consumption rates in the world.[26]

Several steps were taken to manage wartime supplies of butter to meet domestic needs and overseas commitments. One of the first actions of the newly created Wartime Prices and Trade Board (WTPB) was to fix the price of butter, to cope with the continuing heavy demand. The Board's equitable distribution plan, which began in October, 1942, required orders of scarce goods to be filled by manufacturers and wholesalers in proportion to 1941 sales to these customers. Butter remained in this program until the end of the war.[27] On December 21, 1942, butter became the first domestically produced food to be rationed; sugar, tea and coffee had been rationed earlier that same year. The WPTB responded to demands developed over the previous half year by such organizations as the Retail Merchants Association,[28] and the Toronto City Council, which even considered its own rationing program.[29] The shortage generated illegal means to acquire butter. One woman used a fake doctor's prescription to obtain a half pound of butter.[30] The original amount stood at 8 ounces per person per week. From January to March, 1943, a temporary drop to 1/3 pound per week was imposed. In March, 1944, the ration was effectively dropped to 7 ounces per week by postponing the validity of one coupon every eight weeks. In December, the ration dropped to 6 ounces.[31]

Buttered popcorn was ruled a luxury for the duration of the war.[32] Home economics classes in high school went without butter.[33] The situation gave Canadian consumers an opportunity to demonstrate their ingenuity. Whipping one pint of warm milk into one pound of softened butter resulted in the equivalent of two pounds of butter.[34] In the Maritimes, a recipe calling for butter, milk, gelatine and colouring, warmed and whipped to the consistency of whipped cream, expanded one pound of butter to 1 1/2 pounds and twice the volume of bread spread.[35]

Inevitably, the possible use of margarine as a diet supplement was discussed. Already, as part of pre-war discussions about food requirements in war, the point had been made that margarine might have to be put into the marketplace again.[36] In reaction to the 1942 rationing of butter, Senator A.C. Hardy, a dairy farmer from Brockville, Ontario, advocated the use of margarine for several reasons: there was a shortfall in the production of butter; the butter that was available had been distributed unevenly throughout the country; the WPTB had bungled its responsibility of encouraging production to meet demand; and the general public was guilty of hoarding whatever butter they could purchase.[37] Senator Hardy generated a surprising amount of support in the national press for the butter lobby's argument that butter was in sufficient supply to meet national needs if proper planning were employed.[38] As well, positive reaction to Hardy's effort came in February, 1943 when the Brantford, Ontario, City Council urged the federal government to allow both the manufacture and importation of margarine.[39]

The butter industry continued to expound a narrow protective position. The B.C. Federation of Agriculture bluntly called for a continuation of the margarine ban.[40] Others called for higher tariffs on vegetable oils as a means of keeping butter's competition in the area of cooking under control.[41]

It may seem strange that what happened regarding margarine in World War I was not repeated in World War II. But there was a key difference in the government's handling of the whole problem of food. In 1914-18, Ottawa had stressed production; when butter producers could not meet demand, margarine was introduced to fill the gap. During World War II, on the other hand, when it became clear that the supply of butter was insufficient, bureaucrats saw rationing as the logical solution. The oils needed to produce margarine were also in short supply.

Margarine, however, did become part of Canadian wartime experiences, for the servicemen sent overseas and for the brides who came to Canada after the war. Margarine was with the Canadian soldiers at Dieppe in 1942.[42] These young men and women were to form a sizeable portion of the adult population and would help in deciding whether margarine should once again be legalized.

It is surely human nature to need something about which to complain. Traditionally soldiers have been among the worst complainers, one of their chief targets being army food. In the Canadian armed forces, butter was the only bread spread given a place in the rations, at least while stationed in Canada. However, once overseas, imperial ration scales, with a Canadian supplement, were put in place. In these scales butter or margarine was included, depending on supply.

In January, 1942, the Standing Committee on Nutrition revamped the military ration to meet recently set nutritional standards. Twelve months later, the ration was reduced by 1/8 to comply with the rationing being imposed on Canadian civilians, a pattern maintained for the rest of the war.[43]

Farley Mowat's memoir, *And No Birds Sang*, offered his opinion of the contents of one of the imperial ration scales.[44] Margarine received its share of caustic comments. "Canned yellow wax" was in the category of the inedibles which were traded to the Italian civilians, whom Mowat suspected of using it for mule feed.[45] Bing Coughlin, the very popular military cartoonist, presented a sample menu (shown on the following page) with the intent of demonstrating the negative reaction margarine generated among soldiers.

Other persons' memories include being so hungry that anything edible was devoured with no difference established as to whether butter or margarine had been served.[46] The tropical variety—there was another for colder climates—was dark yellow in colour, with a strong taste. It did not spread, but was eaten crumbled on bread.[47] Ross Munro, the war correspondent, said he often "ditched [his] lunch [of haversack rations] to avoid the margarine. On the other hand, a tin of butter from home or one found in remote areas when on leave, was treated like gold."[48] The disinclination of the men to eat margarine and an inability to trade all of it to the Italian civilians, generated a backlog of supplies. Excess margarine and the desire for a party led to using margarine to make excellent French-fried potatoes.[49]

Among prisoners of war, butter easily came out ahead of margarine on the list of special treats. The Canadian Red Cross packed 16 million food parcels which were sent to Canadian, British and Allied prisoners of war. The goal was to provide one parcel per week to each prisoner. The program was evaluated over the winter of 1943-44 by a number of Canadian Army and R.C.A.F. personnel who had been repatriated from Germany or escaped from Italy, and again by 6,551 Canadian and British servicemen when they were released from prisoner of war camps in early 1945. The German ration scale had provided a total fat issue of 4 ounces of margarine per week. This major deficiency was supplemented by one pound of butter in the Canadian Red Cross parcel. The most appreciated food was a white-flour biscuit: butter was second.

Margarine was never included in the Canadian Red Cross parcels, but the prisoners of war received British and American parcels, with either butter or margarine, about 70% of the time. Therefore, Canadian personnel had a very

BREAKFAST
Egg Cosmetic
Bacon Limp
Toast Margarined
Tea

LUNCH
M & V (Stew Dehydrated)
Potatoes Dehydrated
Prunes
Bread Demoulded
Tea

DINNER
Soup Strained
Choice of: Bully or Mutton Dehydrated
Vegetables (See Lunch)
Pudding, Rice Raisinless
Biscuits, Tack Hard
Tea

"BROKE IT TRYIN' TO SPREAD THIS NEW KINDA MARGARINE!"

Source: Bing Coughlin, *Herbie* (Toronto: Thomas Nelson, n.d.), 122.

good chance of having received margarine at some time during their incarceration. Butter won the competition almost unanimously. It was butter that made Canadian parcels the preferred choice:[50] one Canadian parcel was equal to two British for trading purposes.[51] It is worth noting that lamps constructed to provide light for the digging of escape tunnels used pajama-cord wicks soaked in German margarine.[52]

The almost half-million men who served overseas probably came away from their military experiences with a greater appreciation of butter than they had when they left home. The extent to which service personnel and civilians were unable to make a distinction between butter and margarine speaks more of the poor quality of the butter being served than of any high quality of the margarine.[53]

For 48,000 war brides, "marge" had been a part of the British diet long before 1939, its place reflecting the class structure. For instance, a middle-class English home almost always had butter on the table but margarine was almost always used for cooking. During the war "marge" worked its way onto the table.[54] For the poorer classes margarine remained the normal spread; butter, if at all available, was a special Sunday treat.[55] The practice of mixing the butter and margarine rations together into one spread seems to have been fairly common.[56] The pre-war experience of a Dutch war bride was the use of a margarine of reasonable quality and cheaper price for cooking. During the war a special margarine was manufactured for the civilian population which was said to taste like "melted candles."[57] The consensus of this group of women seems to have been that unless economic circumstances dictated otherwise, butter would certainly be used in their homes as the bread spread and, if possible, in the kitchen as well.

The proponents of butter were able to maintain the ban on margarine in the inter-war and war years. It would require the post-war desire for a higher standard of living to generate the demand to legalize margarine, because not enough butter could be produced.

PART TWO
The End of the Ban

CHAPTER SIX

The Fight to End the Ban 1946-48

An urbanized Canadian society emerged from the Depression of the 1930s determined that such suffering should never happen again. The experience of World War II demonstrated to the generation now ready to assume leadership, that people could use their knowledge, especially in the social sciences, to alter the social order for society's benefit. The immediate goal had been the winning of the war: the long-range goal was the winning of the peace through the achievement of a more prosperous material life for all citizens. For the urban Canadian, one manifestation of this drive for security and stability would be the desire to achieve and maintain a proper nutritional standard. Once there was insufficient butter to meet demand, the cry for margarine was raised. The dairy farmer fought back to protect his own concept of security and stability, the loss of which he feared even more than the loss of income.[1]

Significantly more urbanized Canadians enjoyed an increasingly high standard of living in this war-dominated decade. As a result of the highest birth rate at that time of any industrialized nation[2] and a large influx of new Canadians, the total population grew by 2.5 million. Record high rates of marriage and family formation had their greatest impact in the urban sector. By the end of the decade, urban families outnumbered their rural counterparts by 2 to 1.[3] General prosperity also increased significantly. The index of average wage rates doubled,[4] and the employment rate moved to above 95%. The majority of Canadians reaching adulthood in this decade were raised in an urban setting and ethos. Their response to the challenge of adulthood would be drawn from the urban context.[5]

Milk producers approached peacetime with some trepidation. They wanted more stable, planned production, so that knowledge of cost and price would allow them to work out profit. These rural entrepreneurs were repeating the arguments of their rural and urban counterparts three-quarters of a century earlier, but they were working within a disastrous global food situation. The war had greatly disturbed world food production patterns. After 1945, the continuing impact of the war as well as poor growing conditions in various parts of the world caused a mounting, generalized food shortage. Drought on five continents in 1946 left 30 million people living at various levels of starvation.[6] With world population increasing by 15-20 million annually, the picture looked grim. In the autumn of 1948, the Canadian Bank of Commerce forecast a continuing, "almost universal" shortage of dairy products for the next several years. The crisis was caused by severe wartime losses in numbers of cattle, drastic curtailment in farm manpower, inadequate supplies of fertilizer and the general disruption of agriculture.[7] During the immediate post-war years, Canadian agriculture, particularly the dairy element, was functioning in a climate of significant insecurity and instability. The *Farmer's Advocate* asked a very sound question: "What Will People Eat Tomorrow?"[8] To gain a proper perspective, however, those same years need to be seen as a period of continuing challenge: what was needed was expansion, not curtailment, of food supplies.

The *Advocate*'s point seems to have been missed by Canadian butter producers. With the cessation of hostilities, production actually shrank. Eastern dairy enterprises had been meeting Canadian export commitments, while western dairy production was utilized to meet domestic needs. A "very cold, backward" spring of 1945 in western Canada, added to generally unsettled conditions, and resulted in a 15-million-pound reduction in butter production. The shortfall forced the federal government to reduce domestic rations from 8 to 6 ounces per week, beginning January 3, 1946.[9] Continuing poor weather led to a further reduction of butter and cheese production in the spring of 1946.[10] Butter supply became so tight that people were considering themselves lucky to be able to buy even one pound of butter for a one-month period. Hotel patrons suffered through butterless days.[11]

Conditions worsened as the year wore on. Unwilling to reduce the butter ration even further because of the possible political backlash, the King government purchased 5-1/2 million pounds of butter from Australia and New Zealand. This food had originally been allocated to Britain, and the British reluctantly gave it up in exchange for more Canadian bacon. Political backlash there would not be: dairy farmers perceived these imports as aiding their cause by continuing to cultivate consumer preference for butter and counteracting both the growing use of marmalade and the increasing demand for margarine.[12]

Wartime price controls on foodstuffs had helped the consumer but had placed the producer in an increasingly tight profit squeeze. By the spring of 1946, Canadian dairymen were receiving only $2.10 per 100 pounds for milk allocated to butter and cheese. This price was virtually half that being paid to U.S. dairymen, even though it included the federal subsidy.[13] However, the virtual 50% tariff and the margarine ban left the Canadian butter industry with an "absolute" monopoly that was "indefensible."[14]

Under these circumstances Senator W.D. Euler determined to take some action. He had recently travelled to Florida, and the presence of margarine in

American restaurants and food stores had renewed his sense of the unfairness of Canadians not being permitted to use this food. "[W]hile I never knew definitely that I was eating margarine, I am sure it was on our table at times, although I could never tell the difference."[15]

Senator Euler's qualities were such that he became leader of the kind of democratic movement into which the pro-margarine forces were to coalesce. In 1917, Euler had achieved a high degree of personal political respect by winning a Commons seat for North Waterloo as a Laurier Liberal, against very stout opposition by the entrenched forces in the Union party. His forthright reputation led Mackenzie King to overcome his personal dislike, and appoint him Minister of Customs and Excise in September 1926, in order to 'clean house' after it was discovered that customs officers were engaged in smuggling liquor and a wide range of other goods to the United States.[16] The following March, he became Minister of National Revenue and in 1940 moved to the Senate.

Euler introduced his private member's bill on March 27, 1946. In his address, he presented arguments grounded in the practical liberalism that he had expounded during thirty-five years in politics: Canadians ought to have freedom of choice between butter and margarine: each had proven themselves to be wholesome foods. Poorer people in particular needed margarine in order to keep food costs under control while maintaining an adequate nutritional level. In this particular instance, he was primarily concerned with protecting the rights of the consumer, but he was quite prepared to appreciate that the producer might have a case and he was willing to listen. However, history had demonstrated that in Canada, between 1918 and 1923, margarine had not become a serious competitor against butter. The co-existence of the two foods in the rest of the world also showed that the use of margarine in Canada need not bode ill for the butter industry. In other words, the well-being of Canadians could be enhanced without detriment to the dairy sector of the economy.

The political debate that followed cut across party affiliations and brought out strong arguments on either side. Euler attracted the support of several senators whose position centred on consumer protection, while not hurting the dairyman. Senator N.P. Lambert, who had been secretary of the Canadian Council of Agriculture, argued the unorthodox, but sound, point that margarine in the marketplace would really help the dairy industry. If Canadians were able to save by purchasing margarine, they would use these savings to buy dairy products in other forms, such as fluid milk and ice cream. Senator W.M. Aseltine supported Lambert by noting that the price paid to the farmer for butterfat destined for butter manufacture was 21¢ per pound lower than when sold as fluid milk. Senator John Haig sought to demonstrate the increased consumption of fluid milk by pointing out that, when his son returned home from war service, the household increased from three to four persons but milk consumption doubled! Senator P.J. Venoit, a physician, argued the case from the viewpoint of health. As a diabetic, he appreciated the need for a finely balanced diet. For example, a diabetic child needed

2 ounces of butter per day, far more than present supplies allowed. Margarine would take some pressure off butter supplies.

Senator J.J. Bench, Euler's chief colleague and legal advisor in this matter until his sudden death in December 1947,[17] advocated a temporary lift of the ban on margarine. His amendment calling for a two-year relaxation of the ban was ruled out of order. Yet, in attempting this manoeuvre, Bench also raised a point that would come to the fore two years later: the question of parliamentary jurisdiction.

Wishart Robertson, government Senate leader, quickly made it clear that the King ministry wanted to avoid arousing the ire of the dairy industry. On a more practical level he also argued that, since world fat supplies were only about 60% of requirements, it would be impossible to find the vegetable oils necessary to produce the margarine. Against Euler's bill, he also brought out the well-worn rural myth: if society was to encourage young Canadians to "return to the land," then all forms of agriculture needed protection and this bill worked in the other direction. He also scored a few economic points: the recent sharp increase in demand for fluid milk put heavy pressure on the dairy industry. This fluid milk market was attractive to the producer because of higher profits. What was needed was an increase in the price for milk destined for the butter industry, to make such production more attractive. At best, a temporary relaxation of the ban might be tolerated, to help meet the emergency created by the contemporary shortage of butter, as had been the case in 1917.

On second reading Euler's bill was defeated 43-30, but he felt reasonably satisfied with the result of his effort. He perceived that no sound argument had been made against his bill, that solid arguments had been made on its behalf, and that leaders of both parties had withheld their support primarily because of the embarrassment which Senate approval would generate in the Commons.[18]

Across Canada, Euler's efforts generated both positive and negative reactions. Press coverage of the debate was general and more sensitive than the politicians in its response to the issue of the public's need. It performed two major services, by providing a forum for public debate and a medium of public education.[19] For the most part, standard reports of the debates were printed: only the Montreal *Gazette* gave the topic front page attention — on April 11 and May 9, 1946. No editorial reaction was generated except by the *Kitchener Daily Record*, of which Euler was part-owner and president.[20] News reports of Euler's action set off an extensive correspondence by private citizens to the Senator, all but one letter being supportive. Several writers had shared his experience of having had margarine while on vacation in the United States, an example of contact with the product that was to help the pro-margarine forces in Canada.

Another factor helping Euler was the European experience of some war brides and veterans. One war bride confided to Euler her desire to continue using margarine in Canada, as she had in Britain. Forty years after the war, other war brides were also to recall similar feelings, although these had not been voiced back in the 1940s. A few veterans also demonstrated support for margarine as a result of their wartime experiences.[21]

The Senate debate generated powerful negative reactions within the dairy industry. The Central Alberta Dairy Pool registered its opposition in a letter to Euler on April 25.[22] The Saskatchewan Dairy Association made its opinions known in letters to Senators and MPs from Saskatchewan and to James Gardiner, Minister of Agriculture.[23] While the urban press was almost unanimous in its support of the legalization of margarine, the agricultural journals supported the dairy side of the issue. "If Canadian Senators must have oleo, for health reasons, it could be manufactured under strict controls and supplied to the Senators on doctor's prescription."[24] This statement illustrates the fears of the butter industry at a time when the fortune of "no other farm product is so charged with political dynamite and the germs of social upheaval."[25] The next battle of the war had now to be fought.

Conflict erupted again on the senatorial front in February, 1947, when Euler reintroduced his repeal legislation. The debate generated two new arguments. Were the ban to remain in force, the hand of the Canadians in Geneva, who were negotiating what was to become the General Agreement on Tariffs and Trade (GATT), would be weakened. Euler argued that the Geneva agreements would not permit prohibitions and therefore his bill only did what would soon be required. Secondly, the negotiations for the entry of Newfoundland into Confederation would be greatly facilitated if the negotiating parties had a common policy on the legality of margarine.[26] The bill was again voted down 38-22. The sensitivity of the topic for the King government becomes evident when it is noted that the issue split both major parties: 11 PCs and 27 Liberals voted against Euler's bill, 3 PCs and 19 Liberals in favour.[27]

Extra-parliamentary activity continued to mount. The urban daily press gave the Senate debate low-key, back-page coverage. Euler continued to receive nearly unanimous support. Encouraged by hundreds of letters from ordinary Canadians, he moved in two directions. Firstly, he requested Senator J.J. Bench, his legal counsel, to institute a civil action to test the validity of the ban.[28] Secondly, with Bench and Alfred Altmann of the National Dairyman's Association of New York City, he began to put together a company to manufacture margarine once the legalization fight had been won. The move seems to have been made in order to help force a test of the ban in the courts, and the project was put aside when Bench died.[29] Euler had attempted, early in 1946, to ascertain under what conditions companies such as Canada Packers and Swift Canadian could begin production, should margarine be legalized, and how Kraft Foods of Chicago would handle the export trade to Canada. Through letters to urban daily newspaper editors, he attempted to convey the reasons for his position to a wide public.[30] The debate also expanded to a more formal campaign in which urban-oriented organizations took a position on the issue. Churches, labour unions, boards of trade, municipal councils, hospitals, veterans and women—all threw their support behind Euler, an action he well appreciated.[31] The fight to legalize margarine was to be one of the first projects of the Consumers' Association of Canada, newly formed in late summer, 1947.[32] The Gallup Poll showed Canadians shifting to an increasing appreciation of margarine. Those supporting margarine's sale had increased, since mid-war years, from a third to over half of those sampled. Margarine gained converts even among the farmers.[33]

The Canadian Medical Association gave the political advocates of margarine great encouragement through an editorial in the mid-summer issue of its *Journal*. The editorial began its argument with the words: "Margarine is a substitute for butter." It concluded with the blunt statement: "From the economic and nutritional aspects good margarine is superior to butter."[34] The pronouncement of the medical profession received generous coverage, even editorially, in the press, each paper condemning or praising as its original position on the issue dictated.[35] The fears of the pro-butter people were thus further heightened. It was vital to fight this perceived threat to their livelihood.

Not all urban elements in the nation had been won to the cause of margarine. The Canadian Chamber of Commerce gave up attempting to reach a position on the issue when a referendum of its membership showed less than two-thirds of the 235 member units were in favour of pressing for federal government action to legalize margarine.[36] Percy Bengough, president of the Canadian Congress of Labour, reported CCL membership was deeply divided on the issue and it would therefore be virtually impossible to arrive at an official position. Pat Conroy, the secretary-treasurer of the CCL, also declined to give his personal support. As a youth in Britain, he had been inculcated with a prejudice against margarine — a "class food," a "poor man's" food. Conroy believed the real effort should aim to get workers an adequate wage to enable them to buy the best food possible, including, of course, butter.[37]

Euler very confidently attacked the argument of his opponents as introducing "irrelevancies," arousing "prejudices" and evoking "ridicule."[38] However, from the standpoint of most dairy farmers and, in particular, the political leadership of the lobbyists, the industry could be provided with a sound defence. The matter was made more urgent when government controls such as rationing ended in June. Butter prices then rose from 53¢ per pound wholesale in early July to 66¢ by the end of November,[39] while consumption increased by 18.7 million pounds.[40] A major debate developed in the butter industry over whether to push profits on volume and thus keep price under control, or to permit prices to climb even higher. The fear of the second group was that a rising price of butter would give cause for margarine to be allowed in the marketplace.

The customary tools of lobbying were expanded. The Dairy Farmers of Canada worked with other farm-related associations to solicit petitions from the grass roots.[41] Members of Parliament and Senators were contacted, and the Prime Minister's office received, in one two-week period, 250 petitions, chiefly from Alberta farmers.[42] Full-page advertisements were purchased in the daily press in an attempt to reach the urban public.[43]

A pamphlet entitled "The Margarine Question" presented the position argued by H.H. Hannam, president of the Canadian Federation of Agriculture, in a debate with Euler in Montreal sponsored by *L'Association Canadienne des Electrices*. The four-page pamphlet, intended for the widest coverage possible, presented the dairy position as succinctly as it could be put. The dairy industry was one of the most important in Canada, involving half of the farm families in the country. Its economic well-being had wide repercussions throughout the nation. Deliberate government policy had kept the price of milk and its byproducts low during World War II. Since 1939, the highest price, including government subsidies, received by the primary producer for a quart of raw milk going into butter had been 4¢ in May, 1947, only twice the low price received at the depth of the Depression. The intent of the wartime policy had been to force as much milk as possible into cheese production to meet, in particular, Britain's wartime needs. Now, in the midst of post-war prosperity, the dairy farmer was being denied his chance to reap the rewards of his earlier sacrifices.

Hannam's pamphlet presented figures which showed that total annual milk production had increased by 1 1/2 billion pounds over pre-war production, and for 1947 the goal of the industry was to add a further half billion pounds. Historically, butter prices had remained low. Since 1914, the average yearly wholesale price had been less than 40¢ per pound for 28 of the 33 years, and less than 30¢ per pound for 10 of those years. The price had been attractive enough to cause the Canadian consumption rate to stand behind only that of New Zealand and Australia. Hannam refuted Euler's point that margarine would sell at a price half that of butter. Many of the same oils which would be used in margarine were being used to produce shortening, which was selling at a retail price of 31¢ per pound. If government vegetable oil subsidies were removed, the price of shortening would rise to 45¢. It was known that margarine cost more to manufacture than shortening. Hannam estimated that margarine would retail at least at 50¢ per pound without government subsidies. Americans were no better off, even though they had two bread spreads from which to choose. American per-capita consumption of butter and margarine was half that of Canada's consumption of butter alone. If all other oils and fats were included, the American total was still a little short of that of the Canadian total. Hannam concluded that the butter people were only seeking the same consideration for protection through government regulation that most Canadian industry was now enjoying. Euler's contention that citizens' rights were being denied through the continuance of the ban on margarine had another side: even with the ban, Canadian rural folk were being denied the right to the same living standard as their urban counterparts enjoyed.

Basically, Hannam had a sound set of arguments: one fundamental point was, however, being missed. The real choice for Canadians, as Euler and the *Globe and Mail*[44] pointed out, was between butter and nothing. Canada was the only country in the world with an absolute ban on margarine. No other product, nutritional or industrial, had such absolute protection from competition.

While Hannam placed himself in a relatively sound position, some of his allies did not. For example, the editor of the *Canadian Dairy and Ice Cream Journal* attempted to counter the argument that margarine would help bring down the cost of living. He pointed out that in the United States and Newfoundland, where the two bread spreads competed, butter sold for more than twice the price of margarine. The inference he drew was that margarine need not be brought onto the Canadian market because butter was the public's proven preference.[45] He failed to recognize the more logical conclusion to be drawn from his own evidence — that margarine in the marketplace would not significantly affect the price of butter, and, therefore, could be introduced without damage to the butter industry.

The third legislative effort by Euler took place in the context of the heightened national economic crisis which had begun during 1947.[46] The most pertinent aspect of that crisis was the rapidly rising cost of living, with its concomitant impact on Canadian attitudes. The "heaven on earth" which the White Paper on Unemployment and Income of 1945 was supposed to have inaugurated was perceived to be in jeopardy. A 9.3% increase in the cost of living index for 1947 was compounded by a further 14.4% increase in 1948, the highest since 1920. These rates of increase pushed the consumer price index to a new high of 156.[47] The price of milk and milk products increased even more, 19.1% in 1947 and 16.4% in 1948.[48]

Graph 6-1
Butter Retail Price Index
(August 1939 = 100)

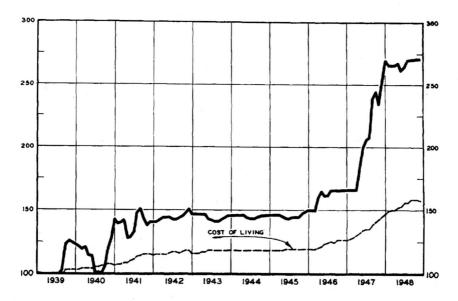

Source: *Report* of Royal Commission on Prices (1949), 72.

Yet such increases were not enough to reverse the annual decline in production since 1945. This shrinkage ran parallel to a similar, nearly steady decline in cow herd size. The smaller amount of milk produced demonstrated its effect in the production figures for butter. From a 1943 peak of 7.3 million pounds of creamery butter, production had dropped 6.7% in 1947, and in 1948 a further 1.8%.[49]

The impact of the decline in supply was itself reflected in the drop in consumption of dairy products in general. From a high in 1942 of 14.3 million pounds, consumption dipped for two years and then rose to a new peak in 1947 of 14.5 million pounds. For creamery butter alone, 1942 had seen a high of 304.7 million pounds. Both 1947 and 1948 were expansion years in total consumption and were exceptions to the fundamental trend. The per capita statistics illustrate the point more clearly. In 1945, an all-time high of 1,244 pounds of milk consumed in all forms had been followed by years of steady decline. For butter, consumption reached a per capita high in 1942 of 751 pounds. The butter consumption statistics ran an unsteady decline through the next four decades.[50]

The inflationary spiral had its greatest impact on the price of individual items. Between September 1947 and January 1948, the increase in the price of butter constituted the largest item in the total cost of a selected list of foods.[51] More broadly speaking, from a Depression low in 1932 of 22.6¢ per pound wholesale price for first grade creamery butter in Montreal, there was an almost steady increase to 40.8¢ in 1946. The price climbed to 52.6¢ in 1947 and soared to 69.8¢ in 1948.[52] Concurrently, the average weekly wage ($36.58) fell short of meeting the cost of basic minimum needs ($40.11).[53] Food alone now cost more than had food, fuel and rent combined immediately prior to the war.[54] That the use of margarine instead of butter could reduce a family's food bill by $15-$20 annually[55] was sufficient motivation to consider shifting, were it possible.

Consumer resistance to increasing prices developed: 55% of families reduced their purchases of butter; 21% bought less bread, 34% fewer eggs, and 51% less milk.[56] Canadians living near the U.S. border were no more fortunate as concerned this alternate source of supply since American food prices were significantly higher.[57] The situation was generating new ideas on how to cut butter consumption, as had been done during the war.[58] The Aylmer Company advertised the use of its jams and marmalade on toast without butter.[59] A drug store and a jewellery store in Montreal offered butter as a bonus on other purchases until consumer objections forced the Wartime Prices and Trade Board to end the practice.[60]

The federal government responded paradoxically to the growing crisis. On the one hand, in order to meet the projected Canadian contributions of foodstuffs, raw materials and manufactured goods to the recently created European Recovery Program (Marshall Plan), government officials appreciated that Canadians would have to cut domestic consumption. The rapid price rise was seen as helping to reduce demand.[61] On the other hand, the domestic political scene required some alleviation of the rising cost of living.

In mid-January, the government reinstated price ceilings on butter. Two days after prices in Halifax hit a peak of 78¢ per pound, the imposed retail price was set at 74¢ for the Maritimes, 73¢ for the St. Lawrence provinces, 71¢ for Manitoba, 70¢ for Saskatchewan and 72¢ for Alberta and British Columbia.[62]

In the spring of 1948, the establishment of a Commons committee to inquire into the question of rising prices provided a forum for many sectors of society to voice their concerns. Butter's scarcity and its relatively high price received a fair bit of attention. Ross Thatcher, CCF MP for Moose Jaw, Saskatchewan, tentatively suggested legalization of margarine as a relief for the butter-related problems. In a strongly negative response the president of Borden's Co. Ltd., M.D. Warner, not only refused to consider that option, but even declined to utter the name of the despised product.[63] The committee's report led to the appointment of a royal commission, headed by C.A. Curtis of Queen's University, to make a further inquiry.[64]

The complications of short production during the winter season continued into the spring. In summer, discussion turned from a concern about the present shortfall of supply to a possible butter famine during the next winter season, unless the government acted.[65] Action in the form of a one-cent increase[66] seemed ludicrous to the producers who were arguing that if price levels were commensurate with costs, butter should retail at 91¢ a pound.[67]

Given such circumstances, it was almost inevitable that the calls for the legalization of margarine would intensify. The Welfare Council, the Hospital Council, the Chemical Institute, the Ontario Federation of Home and School Associations, the Ontario Firefighters' Federation, the Lumbermen's Association, and the National Council of Women were among the bodies joining the chorus in support of legalization.[68] Debating teams at Kitchener-Waterloo Collegiate Institute used the topic and the victor was not always on the side of pure logic.[69] The Brantford, Ontario, City Council, repeating its 1943 action, passed a resolution supporting margarine's use and called on all city councils to act in a like manner.[70]

The issue was forced on the national political parties preparing for an approaching federal election. For most Canadian politicians, the editorialist for the *Farmer's Advocate* expressed the political situation well: "No political party in Canada will allow itself to be manoeuvred on to skids greased with oleo."[71] James Sinclair's (Liberal, Vancouver North, B.C.) effort to place the topic before the national Liberal convention failed to pass the B.C. caucus meeting.[72] The national Liberal leadership ducked behind the contention that since the question was *sub judice*, discussion and action by the convention was not acceptable.[73] The Progressive Conservatives did not operate under such narrow guidelines. In one of his last public statements as leader, John Bracken supported the butter lobby's position.[74] At this convention, which chose George Drew as new leader, an amendment to a resolution regarding the high cost of living was passed by an overwhelming majority. It called for approval of the "production and sale of all foods beneficial to health." The circuitous language was deliberately employed to avoid the problem of discussing a topic which was then before the courts. The mover of the amendment made it quite clear that he had no food other than margarine in mind.[75] The CCF was much more forthright than the other two parties. Its platform stipulated that the party was in favour of the "importation and manufacture of margarine exclusively by a Crown company, so that the quantity, quality and price may be regulated according to the needs of the dairy industry and the consumer."[76] Why this food should be dealt with in such exclusive fashion was not explained. There is no evidence that the Social Credit party reacted to the issue and no reference was made to margarine in its platform.

The butter forces were demonstrating that the ranks of the industry were a solid phalanx in the struggle against margarine. Under the auspices of H.H. Hannam, President of the Canadian Federation of Agriculture, and Earle Kitchen, Secretary-Treasurer of the Dairy Farmers of Canada, supportive petitions were solicited from agriculturists across the country. The numerical breakdown of returns — B.C. 9, Alberta 11, Saskatchewan 31, Manitoba 2, Ontario 72, Quebec 20, Nova Scotia 12, New Brunswick 22, and P.E.I. 10 — demonstrated again the preponderant interest of the Ontario sector of dairying.[77]

The pressure on MPs and Senators through letters continued unabated.[78] A thirty-page booklet, "Will It Be of Benefit to Canada?" was prepared by the Dairy Farmers of Canada for nationwide circulation.[79] The CBC's Farm Radio Forum was used to spread the message.[80] Gilbert McMillan, President of the Dairy Farmers of Canada, headed a national committee entrusted with the coordination of the struggle. The committee was to collect a special campaign fund (the amount collected is unknown). McMillan assumed full-time status as chief lobbyist so long as legislative proposals were before the Commons and Senate.[81]

In spite of all of this activity, there were doubts in the ranks of the faithful. Some individual dairymen, such as Senator Hardy, had long been arguing that the removal of the ban would not harm the industry. C.H.P. Killick, Dairy Commissioner for Manitoba, was the first bureaucrat connected with dairying to admit being in a quandary and to be able to see some validity in each side's arguments.[82]

The focus of all this attention and activity was parliament. On December 9, 1947, Euler had introduced for a third time his bill to legalize margarine. He was joined in his personal crusade by James Sinclair who introduced a bill, identical in intent, in the House of Commons one week later.[83] Sinclair had joined the fray as a result of a constituent's query at a forum on the high cost of living. Unable to explain to her satisfaction why margarine was not available to Canadians, he determined to help advance the work being done by Euler and others.[84] He had even less success in the Commons than Euler was having in the Senate. Sinclair also worked to help the cause by debating the issue on the Farm Radio Forum.[85]

The bulk of the Senate debate took place in the two weeks after April 20, 1948. It received extensive coverage in the back pages of the daily urban press. A new element was introduced: a similar debate was being waged in the United States, although the focus there was on the colour and tax issues. In February, the New York state legislature reached a compromise by suspending the ban on coloured margarine for one year.[86] In April, the colour ban was lifted in New Jersey.[87]

The high point of the Senate debate occurred on April 28, when the government leader, Wishart Robertson, took 1 1/2 hours to reiterate what he labelled as his "personal" views, even though he had cleared these with the cabinet. During his address, he took no definite position as to whether the ban was right or wrong in principle. Rather he contented himself with expounding on the need to protect the well-being of the dairy industry. He believed it "cruel deception" to attempt to have Canadian consumers believe that the edible oil supply situation would make margarine available, were it to be legalized. Senator James Murdoch responded heatedly that he believed Robertson's assertion was "absolutely untrue." In a fit of hyperbole, he accused the Speaker of using "communist tactics" to stop margarine's supporters from gaining satisfaction. Senator Paterson probably came closest to the crux of the issue from a common sense point of view: the wishes of 150,000 producers of milk had to give way to the desires of 13 million consumers. Euler opened and closed the debate, but developed no new argument of substance.[88] The conclusion of the debate coincided with *Halifax Herald* headlines about the cost of living index having reached a new high. The vote in the Senate stood at 35-21 against the bill — 28 Liberals and 7 PC's against, 17 Liberals and 4 PC's in favour. The decision to keep this "glaring example of discriminatory and protectionist legislation" was roundly condemned in the press.[89]

Tactically, two moves arose out of the debate. The government was beginning to face the reality that it could not win, no matter what action it took. Following up a point raised by Senator Ian Mackenzie, Robertson (on May 14) stated that it was now appropriate to consider an inquiry into all aspects of the fats and oil situation in Canada. Time was too short to complete such an inquiry before the present session of parliament drew to a close, but in the next session he would be prepared to move that the Standing Committee on Natural Resources carry out such an inquiry — a foot-dragging operation, pending a solution.[90]

Much more significant was Euler's announcement that he would be reintroducing his bill next year, but that in the meantime he would be following through on the point of constitutionality raised in 1947 by Senator Bench. On June 9, Euler moved, seconded by Senator Hardy, that the government should seek the Supreme Court's opinion regarding the constitutionality of that part of the Dairy Industry Act of 1927 which dealt with the prohibition of the manufacture or sale of margarine in Canada.[91] (The ban on importation was excluded because the federal government's jurisdiction in this area had never been questioned.) The following day Senator Arthur Roebuck presented a synoptic introduction to the legal points to be raised before the

Supreme Court: margarine in the marketplace did not fall within federal jurisdiction under criminal law, trade and commerce, and/or peace, order and good government. It was far more logical to consider that margarine came under property and civil rights, which came under provincial jurisdiction, thus the ban of 1886 was, in the 1940s, *ultra vires* of the federal parliament.[92] The vote was unanimously in favour of the referral procedure. The move also received unanimous approval in the daily urban press, although the rural papers showed some apprehension.

The previous week had seen the demise of the Sinclair bill in the Commons. As long as the margarine issue remained on only the Senate's docket, the King government could avoid any real commitment. But with the issue now before the Commons, the cabinet was required to become directly involved in the conflict.

The first reference to the margarine issue in Mackenzie King's diary occurs on March 19, 1947. Louis St. Laurent, Secretary of State for External Affairs, had made a "splendid" speech to caucus on the need to keep the margarine bill—"certain to divide our party in a manner which would be most harmful"—out of the Commons. King expressed delight that Euler had taken the opportunity to attend caucus to hear St. Laurent's argument and note his positive reception. King referred to Euler as having "asked me previously if he should take the line he did. I told him by all means to do so."[93]

Euler was independent-minded to a degree sufficient to take hold of this issue in spite of an appreciation of its prickly nature for the government. But he was also enough of a team man and practical politician to realize that touching base with his party's leader was good politics and proper courtesy. The real point of interest arises from King's reply. Exactly when Euler made his request is not clear, but it was most likely shortly before he introduced his bill for the first time, early in 1946. Politically, King was an inveterate procrastinator and reference to the courts would postpone the need to make a decision. As well, he must have remembered that, during his first tenure as Prime Minister a quarter century earlier, the question of reimposing the ban had caused serious political dissention in the ministry, and King abhorred disunity. Thirdly, King must also have had bad memories of 1930 when the negative reaction among rural voters to permitting the importation of New Zealand butter helped to defeat his party.[94] For these reasons, to interpret King's statement as a go-ahead in order deliberately to bring an issue guaranteed to disrupt national and party unity to a head, does not fit King's character. The topic had not even been discussed in cabinet since King's third stint as Prime Minister had begun in 1935.[95] King's affirmative response makes far more sense if one sees it as another of King's interminable manoeuvrings. To say "no" to Euler would accomplish nothing.

Nevertheless circumstances were forcing the government to face the margarine issue. The loosening of international trade channels in the postwar era was one of the requirements for the prosperity all desired. The negotiations in Geneva in 1947, which led to the General Agreement on Tariffs and Trade, forced the Canadian government to act. All quantitative restrictions on trade were to be removed. The government had to respond, knowing full well that margarine importation would be affected. Cabinet discussions in February, 1948, centred on two ramifications of the proposed GATT conditions: the feasibility of continuing the ban on domestic manufacture of margarine, and the degree to which a tariff or excise tax replacing the ban might achieve the same result.[96]

Milton Gregg, Veterans Affairs minister, and a veteran of both world wars, let it be known that he had eaten lots of margarine while in the service and was convinced it was not harmful.[97] James Gardiner, from Saskatchewan and Minister of Agriculture, was a staunch supporter of the dairy industry. He consistently expounded the arguments of the dairy interests. In April, 1948, when the government was evidently already taking steps to accede to circumstances, during an interview in Calgary Gardiner speculated that margarine in the marketplace would hurt the dairy industry's capacity to provide an adequate supply of fluid milk. In the *Globe and Mail*'s words, this "specious and devious" argument[98] was typical of his inability to appreciate the complexities of an issue, and his willingness to go to any length to argue the case of his constituents, no matter how partisan. The storm of protest he aroused in the urban press with his inference that margarine was nutritionally inferior caught up with him in the Commons. He quickly acknowledged that margarine and butter were of equal quality.[99] But Gardiner's brashness and King's propensity to evade contentious issues were enough to cause the government not to act on the issue. The pro-margarine case had not yet generated sufficient political heat to force the other cabinet heavyweights into a public commitment.

Mackenzie King had long dreamed of bringing Newfoundland into Confederation. Negotiations for this historic event also brought the margarine issue to the fore. Newfoundland had never placed any restrictions on margarine's manufacture and use. The island negotiating team insisted that this should continue after union with Canada. For the Ottawa negotiators, led by Brooke Claxton, the problem was much more complex. Section 121 of the British North America Act forbade any imposition of barriers to interprovincial trade. This section embodied one of the prime reasons for Confederation in 1867. Section 121, Newfoundland union, and the margarine ban were not compatible. The negotiations regarding union were to continue contemporaneously with the events of the butter/margarine war. The unconstitutional resolution of the issue was incorporated in section 46 of the terms of union whereby Newfoundland, as a province of Canada, was permitted to manufacture and use margarine, but trade of the product between Newfoundland and any other province of Canada was prohibited until such time as trade was made legal between the other provinces. Newfoundland margarine would still be required to meet Canadian standards of quality.[100] Union was promulgated only four days before the Supreme Court was to declare the federal margarine ban *ultra vires*. Newfoundlanders were to boast that the success of the pro-margarine forces owed something to the islanders because of the great desire in Canada for union.[101]

It was evident that historical circumstances were moving the federal government to the acceptance of margarine in the marketplace. In December, 1947, the Department of National Health and Welfare had issued a pamphlet entitled "Canadian Nutritional Notes for December 1947," which concluded that margarine was nutritionally sound. This obvious conclusion is not what drew attention to the pamphlet. The question raised—and left unanswered—was: who had authorized the allocation of time and money for a department of the government to investigate a product which had no legal existence in Canada?[102]

Within this complex of circumstances the King government attempted to deal with what it perceived as a thorny national issue. King did not bend to any pressure within cabinet representing urban, industrial centres, let alone from the country at large. On April 7, St. Laurent, speaking for the absent Prime Minister, gave notice that the time allocated for private members' bills would be reassigned to government business for the rest of the parliamentary session. Every session of parliament had seen such a move by the government. What made this move extraordinary was its timing. To act in this manner so early in a session, and only a week before Sinclair's margarine bill was scheduled for debate, was interpreted by M.J. Coldwell, leader of the CCF, and external observers, as illustrating the government's "sinister" motives.[103] Several private member's bills, in particular Sinclair's, could thus be prevented from creating embarrassing situations for the government.

The government reluctantly backed off. Two weeks later, Sinclair's bill finally reached the floor of the House, coincidentally with the Senate's spending a good portion of its day on the same topic. Sinclair made a superb presentation of his case. He reiterated the arguments raised by Euler and others on previous occasions. As to his own experience, he had eaten either New Zealand butter or British margarine in his wartime rations, while serving in Libya. He had been unable to perceive any difference in taste, colour or texture. He presented to each political party in the Commons arguments couched in terms of its particular basic political philosophy. Progressive Conservatives ought to vote for his bill because of their party's traditional support of national industrial development—this position had been Meighen's when he supported the ban's demise in 1923. To help the poor and to destroy private monopoly were reasons set before the CCF. The Social Credit party's appreciation of free competition required it to support his bill. To his Liberal colleagues he suggested their long-standing appreciation of social justice and equality were sound causes for support.[104]

Sinclair's effort was for naught. The bill was given its hour in the parliamentary spotlight, was talked out, and passed into supposed oblivion.[105] During the debate, George Cruikshank, Liberal member for Fraser Valley, a B.C. riding in which dairying was an important activity, brought into the House a red-ribboned package and gave it to Sinclair. The white contents, supposedly margarine, were passed around the House. It was only after this attempt at some humour that Cruikshank admitted the substance was, in reality, axle grease.[106] This self-appointed whip of the anti-margarine forces in the Commons had prepared a line-up of Liberal MPs who were ready to filibuster Sinclair's bill. "I intend to wind up the debate myself along about 1954."[107]

In early July, 1948, the tremendous pressure building up against the butter interests began to show its effect. The first indication of a shift in the butter interests' position came when the Ontario Cheese Producers Marketing Board resolved to accept margarine in the Canadian marketplace, but on the condition that the export market for dairy products, in particular to the United States, be opened wide and that dairy farmers be permitted to import production essentials.[108] This proposal was forwarded to the national office of the Dairy Farmers. It was based on positive thinking: with the world food shortage to cope with, the need was for greater production of all foodstuffs. The Canadian dairy industry could quite readily have joined this vast productive effort under these new guidelines. Only one other change had to be made. Butter was the only item under a price ceiling. The public feared the removal of the ceiling, and an anticipated escalation of price, at a time when price was perceived to have placed butter out of the reach of many citizens. However, price competition in a market subject to world conditions, as well as the presence of margarine, might have led to steadier and even lower prices, as the Canadian butter producers would have found themselves in a more stable and, if they were more efficient, more profitable situation.[109] As it was, in midsummer, the peak production period, butter in storage was 13 million pounds less than a year earlier. The quantity was 10 million pounds short of requirements for the approaching winter season. Yet, in the first half of the year, 50,000 Canadian dairy cattle had been exported to the United States; these cows could have produced the milk equivalent of 10 million pounds of butter, thus leaving the country in a balanced rather than a deficit position.

On August 10, the National Dairy Council of Canada met with representatives of the federal cabinet. In its submission the Council pointed out a growth in Canadian population of 13% since 1940. In that same period milk production had expanded only 3%.[110] Butter was the one milk product in short supply even though 45-50% of milk production went into butter.[111] What was needed was a freeing of the marketplace for butter: at the present ceiling price, the producer received only 68 cents for the milk used to make one pound of butter. Were that same amount of milk sold in fluid form, 95 cents would be returned to the producer. In order to deal with the short supply, the Council offered four recommendations:

1. importation of adequate butter supplies;

2. in the event importation proved impossible, reinstating rationing;

3. in the event importation and rationing were not feasible "the sale of a suitable substitute product should be authorized until such time as the butter supply is sufficient to meet the demand";

4. in the event of employing a substitute the ceiling price of butter should be removed to permit free market forces to determine butter's position in the marketplace.[112]

Ontario cheesemakers and the National Dairy Council had not mentioned margarine by name. They had noted that the production of an alternate dairy spread, made from butter and powdered milk, should be investigated. Importation in a fat-short world could not be guaranteed, and rationing of one product was not feasible.

Cabinet's response to the dairymen was courteously perfunctory. The decision had already been made to refer the ban's legality to the Supreme Court. In late September, the government succeeded in negotiating the purchase of 15,680,000 pounds of butter from New Zealand, Denmark, and Australia. The step was not perceived as resolving the problem, for price as much as availability determined whether Canadians would have butter on the table.[113] The Gallup Poll taken just before the purchase was announced illustrated citizens' concern: 68% favoured the legalization of margarine, and reduction of the cost of living was overwhelmingly cited as the cause.[114]

On the previous day, the Supreme Court had begun to hear arguments related to the validity of the ban. Political, economic and social factors coalesced to make the entrance of margarine into the marketplace a logical step. As the way out of its perceived dilemma the King government, declining to act sensibly, let alone boldly, opted to let the Supreme Court take on an essentially political issue.

CHAPTER SEVEN

Reference to the Courts 1949-50

Unwilling to resolve whether to meet the desires of 400,000 dairy farmers and their families or the wishes of 13 million consumers, the King government fell eagerly upon a referral to the courts as an alternative. The courts determined the issue primarily on legal grounds, although paying close attention to social causes. As a result the Supreme Court in Ottawa and the Judicial Committee of the Privy Council (JCPC) in London held the ban on margarine production *ultra vires* of parliament. The freedom of choice of Canadian consumers was to be protected. The dairy farmers' belief that margarine in the marketplace would be a threat to their livelihood was to prove unfounded, except in the short run.[1]

Senator John Bench had first publicly raised the issue of constitutionality during the initial Senate debate on the subject. Senator Euler's action, two years later, to move a resolution requesting the government to refer the issue to the courts, was politically motivated. He wanted margarine in the marketplace and was prepared to take any proper steps necessary to achieve that goal: he did not care about establishing any further delineation of federal/provincial powers within sections 91 and 92 of the British North America Act, 1867. Had Euler held no hope of winning before the courts (or, hypothetically, had all nine provincial ministries announced that they would use any new authority gained from a court decision to enact a provincial version of the ban), he would not have used this legal tactic. The Prime Minister was prepared to let some other body make the choice for him, one normally beyond the reach of public criticism. The opinion of the Supreme Court is advisory to the governor-in-council, but in practice the federal or provincial governments involved have respected that advice. Some federal political leaders and civil servants must also have hoped for a court decision against the ban, so that the constitutional anomaly in Newfoundland's entry into Confederation would be eliminated.

Much more importantly, the political crisis related to the shortage of butter and its high price across the nation demanded some response. The federal government deliberately promoted judicial review as a means whereby the prohibition of margarine could be struck down on legal grounds. Lastly, referral was also an acceptable political device for the proponents of butter, of margarine, and the general public to the degree they had an interest in the subject. All hoped that their own interests would be served.

Any critical evaluation of the decision by the Supreme Court of Canada in the *Margarine Reference* must take note of the constraints upon the Court. At that time, it still functioned as an intermediate court of appeal: decisions were subject to review by the Judicial Committee of the Privy Council on further appeal. As well, and vital to any assessment of the Court's performance, was the binding or at least the persuasive nature of previous rulings by the Court itself and the JCPC: in other words, the principle of *stare decisis*.

The historical context for the *Margarine Reference* was that since 1896, the JCPC had eroded the power of the federal government and thereby altered the structure of the Canadian constitution. Not only had the British justices greatly restricted federal regulatory power as it related to both sections 91 and 92, they also restricted use of the peace, order and good government clause of section 91 to periods of national emergency. This "emergency doctrine" meant that the effective residual power had been transferred to the provinces in the property and civil rights head (13) of section 92.

On July 27, 1948, the Canadian government issued P.C. 3365,[2] referring to the Supreme Court the question of the constitutional validity of the *Dairy Industry Act*, R.S.C., 1927, cap 45, sec. 5 (a). Pursuant to the rules of the Court, Justice Patrick Kerwin set the date for the hearing and notified the respective attorneys-general of the nine provinces, the Canadian Federation of Agriculture, the National Dairy Council of Canada, the Consumers' Association of Canada, the Canadian Manufacturers Association and Senator Euler. Responses were made by the government of Canada, the Canadian Federation of Agriculture, the province of Quebec, the Consumers' Association of Canada, *L'Association Canadienne des Electrices et autres*, and also W.D. Euler and others. The first two parties appeared in support of the contention that the subject matter of section 5 (a), prohibiting butter substitutes, was valid. The remaining four parties argued that parliament had exceeded its powers in banning margarine and that the subject was exclusively within provincial jurisdiction. Somewhat surprisingly the National Dairy Council was not represented, although it no doubt contributed to the legal costs of the Canadian Federation of Agriculture. The decision had been taken for Canadian agriculture to present a common front by having only the Federation argue the case.

There are two approaches to the task of a review court when it considers the constitutional validity of a law. The first is the analytical-positivist position, that is, the plain meaning of the words: the judge ascertains the meaning of a law from the way the statute is written, unaffected by any external considerations. The second is the "Brandeis approach": the judge looks beyond the letter of the law and takes into account relevant social and political factors. In its constitutional judgements, the Canadian Supreme Court had historically tended to follow the former position; but in the *Margarine Reference* the Court did not follow this norm rigidly. It heard and accepted for consideration a great deal of extrinsic material. The Court must also have been aware that the effect of finding section 5 (a) *ultra vires* would be to legalize the manufacture and sale of margarine across the country.

The Court spent the four days from October 5 to 8 in hearing the various parties expand upon the factums already presented to the Court. The oral submissions received extensive coverage in the urban daily press. The extent of national interest in the issue is also illustrated by the fact that the twenty-eight seats for spectators in the courtroom were filled and an additional fifty persons stood in the aisles.[3]

In each submission there was a mixture of legal argument and extrinsic evidence. Frederick Varcoe, acting for the Attorney-General of Canada, opened a new chapter in the history of margarine. In arguing that section 5 (a) was *intra vires* parliament, he presented four points. First, the national government and those of the provinces divided between them all the governing powers. Any subject not assigned by the BNA Act to the provinces (section 92) lay within the scope of the national government. Since the margarine ban had not been assigned to the provinces, the federal government was left to deal with the subject. Second, previous court decisions had interpreted criminal law, in its widest sense, to mean an act which carried penal consequences. The Dairy Industry Act carried such penalties, and therefore had to be considered to be criminal law, again a federal responsibility. Third, the Court should understand this act to be dealing with commerce, again a federal responsibility. And last, the Act had to be seen as a means to protect agriculture, a subject on which both governments could legislate, but Ottawa had superior authority.

Extrinsic evidence was also submitted by Varcoe. Because of climatic conditions, the dairy industry in Canada suffered excessive costs when compared to the industry in other countries. Stabling herds for half the year generated major expenses. Because the grazing season was short, milk production was in a lopsided cycle. The large summer excess was the basis for the heavy production of butter and cheese throughout Canada. Butter production accounted for about one-third of the income for the dairy industry. Geography also meant that some provinces would have surpluses and others, deficits, resulting in a considerable interprovincial trade. To Varcoe, this demonstrated the point that regulation of the dairy industry was necessary, and was related to trade and commerce, a federal responsibility. His case was not weakened by the fact that section 5 (a) dealt exclusively with margarine, a dairy product substitute. The comments regarding the peculiar high cost situation for the Canadian dairy industry were presented in the factum. Although no attempt was made to support the law under challenge directly, Varcoe must have submitted this material to convince the judges, by a reasoned explanation, that the ban was necessary for the survival of the dairy industry.

The Canadian Federation of Agriculture submitted the only other brief in support of the ban's constitutionality. R.H. Milliken, counsel for the Federation, was so satisfied with the case presented by Varcoe that he took only twenty minutes, contenting himself with showing how some of Varcoe's points related specifically to the dairy farmers.[4] Varcoe, in contrast, had used one and a half days to work through his submission. Breaking the rule against using parliamentary speeches in court to facilitate the interpretation of a statute, Milliken presented two quotations from Hansard, 1886, which described in gruesome detail the supposed obnoxious process of margarine manufacture. The purpose of this presentation is not clear, since the statements were clearly anachronistic. Moreover, such statements surely would not have influenced the Court in the direction of continuing the ban on margarine.

The factum presented by "Euler and others" contained the substance of the arguments found in the four briefs which contended that section 5 (a) was *ultra vires*. Salter Hayden extended his senatorial support by representing Euler before the Court. The legal arguments began with the assertion that the subject did not fall within the jurisdiction of parliament. As well, provincial regulation of certain products did not apply because regulation implied continuing existence, not prohibition. Margarine needed to be considered as an article of commerce and under provincial jurisdiction, rather than an article of agriculture. Furthermore, the issue was not of such a nature as to have the emergency powers of parliament come into effect. Also, the manufacture and use of margarine clearly came under provincial jurisdiction over property and civil rights. Lastly, the issue was not of national relevance, but rather purely provincial and local in all aspects, and therefore could not be said to fall outside section 92. Euler's factum also went beyond the legal argument to that of reasonable explanation. In an effort to counter the position of the Canadian Federation of Agriculture, it emphasized the proposition that margarine was a wholesome product. Sale had been permitted in Canada between 1917 and 1924. During this time no evidence had been found that margarine was a danger to public health; rather, it performed a useful public service. In later revisions of the original Act the reference to the harmful impact on the public's health had even been deleted.

L.E. Beaulieu, the representative of the Attorney-General of Quebec, argued three legal points. The first two reiterated those raised by Euler's factum, i.e., that the issue in dispute related to property and civil rights, and was not a subject for regulation under the federal trade and commerce clause of section 91 of the BNA Act. The third point countered the federal government's second point: the law was not, in pith and substance, criminal law. The inclusion of penalties could not be used to disguise the real purpose of the legislation, which was to help the dairy industry.

Margaret Hyndman represented the Consumers' Association of Canada before the Supreme Court. The CAC argued that section 5 (a) was wholly *ultra vires* of parliament. It repeated Euler's arguments, expanding on the question whether regulation of margarine came under the peace, order and good government clause of section 91. The various dairy acts had not been passed in times of national emergency. In fact the legislation had been suspended by order-in-council under the War Measures Act in the emergency circumstances of World War I. After the fact, *Maclean's* magazine was to credit Hyndman with having the greatest impact of all the lawyers.[5]

In addition to the above-noted factums, non-legal evidence was also submitted to the Supreme Court by the federal government: statistics on the total number of milk cows in each province, as well as the total per 100 persons; the booklet, *The Dairy Situation in Canada*, issued by the Dominion Bureau of Statistics; a table setting out butter production and estimated consumption by province; and lastly, data concerning production, composition and consumption of butter and margarine in 1939 in most of the major countries. Two departments of the federal government provided non-legal data. Agriculture offered detailed information on the history of margarine, its ingredients, and the method of manufacture. From National Health and Welfare came an extract from the article in the *Canadian Medical Association Journal* which amounted to nothing less than a solid endorsement of margarine as a healthful food.

These items are the most obvious examples of submissions that went beyond legal argument. They lend further support to the contention that the Canadian government, through the office of the Attorney-General of Canada, was merely going through the motions of defending the legislation. Once a reference had been decided upon by the government's political leaders, the Attorney-General's office had no choice but to participate. It also had no option but to argue the legality of federal legislation.

Activity by the Departments of Agriculture and of National Health and Welfare demonstrates the true position of the King ministry because it was no more compulsory for these two branches of government to act than was their choice of which side to support. It was not coincidental, nor accidental, that these ministries both came out in support of margarine.

The decision of the Supreme Court was handed down on December 14. It was timed to meet the deadline of the evening urban dailies, and they pushed aside news about civil war in China and the latest moves in the Berlin Blockade crisis to give the decision front-page coverage, a procedure repeated by the morning dailies on the following day.

The Court divided 5 to 2 in favour of the opinion that section 5 (a) of the Dairy Industry Act, 1927, as it related to manufacture and use of margarine in Canada, was *ultra vires* of parliament.[6] The majority of the Court (Taschereau, Rand, Kellock, Estey and Locke) held that prohibition of "manufacture, offer, sale or possession for sale" of margarine was *ultra vires*. These justices placed the margarine ban in the category of legislation relating to property and civil rights and thus under provincial jurisdiction. Chief Justice Rinfret and Justice Kerwin dissented. The Chief Justice argued that section 5 (a) was within the realm of agriculture; that the Act dealt not with prohibition, but regulation of trade and commerce in a broad sense and not only of one product; and that the issue was of national concern. Kerwin joined his fellow justice in dissenting, but only on the grounds that the legislation was a matter of criminal law. The Court, with only Justice Locke dissenting, also held that the prohibition of the importation of margarine was *intra vires* of parliament.

An analysis of the respective positions taken by the Court's members sheds a good deal of light on a variety of subjects. The judgement written by Thibodeau Rinfret was one of the most important from the standpoint of the dairy industry. He was the Chief Justice, had the longest tenure and provided the most complete range of arguments in favour of federal jurisdiction in the field. However, in hindsight, his is the weakest of all the judgements. Rinfret first reviewed, in exhaustive fashion, the background material provided and the legislative history of the margarine ban. He then analyzed, section by section, the Dairy Industry Act of 1927, and the supporting regulations, in order to provide an appreciation of the "exact purport" of section 5 (a). Rinfret concluded that the section was *intra vires* of parliament because it dealt with a matter of agriculture and thus came within parliament's jurisdiction under section 95 of the BNA Act. Margarine, he contended, was a product of agriculture, not an article of trade. Of course it was manufactured but so were butter, cheese and ice cream; the latter two foods were covered by other sections of the Act whose constitutional validity were not being questioned. Surely one might object that some distinction must be made between margarine, in which milk was a relatively minor ingredient, and butter, cheese and ice cream, which were made basically from milk.

Rinfret then wrote: "It was not contended at the bar — and I think it could hardly be contended — that the Dairy Industry Act and regulations thereunder are not within the domain of the federal parliament by force of section 95 of our constitution." It is difficult to fathom why Rinfret made such a statement. The factums submitted by *l'Association Canadienne des Electrices et Autres*, and by Senator Euler specifically contended that margarine was not to be considered as a product of agriculture. The factum of the Attorney-General of Canada did argue that the challenged section fell under section 95: Rinfret accepted this point. But he compounded the confusion by dismissing the statement of one judge in the major precedent presented as having all the "characteristics of a mere *obiter* and which I consider was quite unnecessary for the purpose of the learned judge in that particular case."[7]

Rinfret's second point rested on more logical ground. The section was valid law because it was criminal law since it provided for penal sentences. Therefore, head 27 of section 91 of the BNA Act applied. He referred to Justice Kerwin's judgement as effectively presenting the rationale on this point.

On point three, Rinfret's reasoning is again difficult to follow. He accepted the argument that since the whole Dairy Industry Act dealt with a variety of products, the topic fell within head 2 of section 91 of the BNA Act, by which the federal government could legislate, not for specific items of trade and commerce, but rather for broader categories. As well, the power to regulate included the power to prohibit things not in accord with the regulations. In making his case, the Chief Justice offered an inexplicable statement: "It has not been disputed that the legislation submitted to us deals with trade and commerce." The factums of both *l'Association des Electrices* and the Consumers' Association of Canada had indeed made such a contention.

The Chief Justice's best argument was seriously weakened by the precedents established by the JCPC as to the clarity of the language of the peace, order and good government clause in the BNA Act. He contended that the sale of butter was a matter of national concern. If margarine's presence in the marketplace were to damage butter's position, then parliament had a right to regulate margarine's existence up to and including prohibition. Such action would be based solely on the law. The possible social desirability of protecting the already low living standards of the poor by providing a bread spread cheaper than butter would not come into play.

In a much briefer judgement, Justice Kerwin presented his reason for dissenting on logical, though extremely narrow grounds. The section under question was clearly within the powers of the national government to deal with under criminal law, section 91 (27) of the BNA Act. The fact that the circumstances behind the passage of the original legislation in 1886 had changed, so that the health of the nation was no longer in jeopardy, did not affect the continuing legality of the ban.

Rinfret and Kerwin each accepted the arguments which Varcoe had presented for the Attorney-General of Canada. Each reviewed the social and economic data submitted to the Court, but based their assent to the validity of the ban strictly on legal grounds. Rinfret's judgement was strongly based on law, but weakly presented. He argued as if the whole of the Dairy Industry Act was being challenged; as well, he dealt with tangential matters and failed to take into account non-legal matters which had been raised. Failing memory or poor notetaking are the most probable reasons for this latter deficiency.

Justice Charles Locke found himself alone in his finding that the whole of section 5 (a), that is, importation as well as manufacture and sale, was *ultra vires*. To Locke, the need to protect the health of the nation ceased being applicable in 1903, when the legislation passed that year made no reference to such justification for continuing the ban. He also was not prepared to accept the argument that the substance of the section was criminal law. Locke concluded that neither section 92 (2) of the BNA Act nor the peace, order and good government clause of section 91 applied, because the presence of margarine in the Canadian marketplace was largely a matter of local or private concern. Interference with the property and civil rights of the people of each province was the relevant issue. As to importation, he recognized that this aspect of the issue had not been dealt with fully by the various litigants. Only *l'Association Canadienne des Electrices* had contended that if the purpose of the ban was to encroach upon the rights of the provinces, then even the importation component was *ultra vires*. The importation ban was unconstitutional because it was ancillary to the two main prohibitions.

Justice Roy Kellock wrote the most extensive judgement on the majority side. He first took up the contention by the Attorney-General of Canada that the original enactment had been valid public health legislation, that this action had not been of a temporary nature and that the dropping of the preamble with its reference to the health factor was immaterial. Kellock rejected this contention, submitting that parliament had, subsequent to the dropping of the preamble, permitted manufacture, sale and importation. Thus the true nature of the ban legislation was revealed—to protect the dairy industry. Next, he dealt with the contention that the ban could not be said to be within the authority of a province under section 92 and therefore had to come under federal jurisdiction. The factum of the federal government had contended that because neither a single province nor all of the provinces together could accomplish the purposes of the ban, jurisdiction then rested with parliament. Kellock rejected the precedents as not applying in the way the federal government wanted them to. He concluded that federal interference in this instance was an encroachment on the power of the provinces to regulate on matters concerning civil rights, a provincial responsibility under section 92 (13). As to the regulation of trade and commerce, he argued that the issue at hand went beyond general to particular controls and thus also stepped into the provincial sphere of responsibility set out in section 92 (13). Kellock also rejected the contention regarding criminal law. In his estimation, once the real intent of the legislation had been established as being the promotion of one business by the prohibition of another, then the criminal law aspect ceased to apply. The peace, order and good government clause also could not apply. Precedents set down by the JCPC had established that this power could be invoked only in instances of national emergency, something which the fate of butter and margarine in the Canadian marketplace did not constitute. As to the applicability of the agriculture argument, Kellock regarded margarine as an article of trade, not a product of agriculture. And lastly, he agreed with Locke that importation was inseparable from manufacture and sale. However, parliament had the right to enact, by whatever type of legislation it preferred, laws to raise revenue or to protect Canadian industry. Therefore, he found section 5 (a) *ultra vires* as to manufacture and sale but *intra vires* as to importation.

Justice Willard Estey also wrote an extended opinion. He began by setting forth his judgement that the aspect of section 5 (a) regarding prohibition of manufacture and sale directly interfered with the freedom of companies and individuals to engage in manufacturing and selling food products. Therefore this part of the Dairy Industry Act was related to property and civil rights within the meaning of section 92 (13) and thus *ultra vires* of parliament unless the legislation could be held *intra vires* via some other provision of the BNA Act. He did not think that section 91 (2), regulation of trade and commerce, applied. He believed that the federal government could only deal with legislation on general, not particular, subjects. As to section 91 (27), criminal law, precedent had established that mere inclusion of criminal penalties in a particular piece of legislation did not make it "criminal." Thus, in Estey's opinion, the substance of section 5 (a) was not criminal law. As to whether the peace, order and good government clause applied, Estey held that it did not. Binding precedent had already stated that in any federal system, it could be envisaged that some beneficial scheme of law might be beyond the power of either jurisdiction. But in this instance the legislation under scrutiny was not a new field which needed to be assigned to one level of government or the other. It was merely an attempt at prohibition of the exercise of civil rights in the provinces and as such, invalid. Estey went on to point out that the problems of the dairy industry, important as they might be as the data demonstrated, had not reached such dimensions as to be a matter of national concern. In several instances, the rest of the Dairy Industry Act had been acknowledged as sound public health legislation. He believed that since margarine was not injurious to public health, section 5 (a) could not be seen as valid public health legislation. As to the applicability of section 95, agriculture, the ban was related to competition of products, not interference with farmers in their agricultural operations.

Justice Robert Taschereau presented the last of the extended opinions. He argued that the problems of the butter industry did not attain the level of importance of a national emergency as precedent had stipulated. Therefore the peace, order and good government clause of section 91 of the BNA Act could not be invoked. Nor did protection of public health come into play. The primary purpose of section 5 (a) was to give butter a preferential position in the Canadian marketplace. As well, precedent had established that parliament could not carve out for itself a sphere of action simply by including penal sanctions in a proposed law. Therefore section 95 did not come into effect because this legislation dealt in its fullest substance with a topic related to property and civil rights. He also accepted the contention that margarine was an article created essentially by a manufacturing process rather than by the activities of a farmer. Therefore section 95 could not be used to raise the matter of constitutionality. And, lastly, he accepted the idea of margarine control coming under the particularity of section 92 (13) rather than the generality of section 91 (2).

Justice Ivan Rand was the last of the justices on the majority side of this case. He concentrated first of all upon the purpose of the legislation related to margarine since the first law had been passed in 1886. This could not be established merely by looking at the preamble of the legislation — actually the original preamble had been deleted in later enactments. He came to the conclusion the ordinary goals of criminal law such as public peace, order, security, health or morality were not served by the margarine ban. The only perceptible purpose was the protection of the production and sale of butter. Therefore, the ban benefited one group against another who would, under normal circumstances, be competitors. Such legislation was an interference with civil rights, a subject listed in section 92 (13) of the BNA Act as under provincial jurisdiction. He joined his fellow justices in their conclusions regarding trade and commerce; peace, order and good government; and agriculture. To Rand, the key point upon which to base his whole opinion was that the thrust of the Act was not the ordinary purposes of criminal law. Rather, the aim of this law was economic: the granting of trade protection to the dairy industry. To do so was to deal with the civil rights of the citizenry, a provincial rather than federal field of jurisdiction.

The single most important factor in the determining of the case by the Supreme Court of Canada was the principle of *stare decisis*. All justices took note of the social and economic data presented to the Court, to varying degrees. However, the decision was strengthened, but not determined, by their use of this material. Repeatedly, justices referred to the precedents set by earlier decisions of the Judicial Committee of the Privy Council. Given the large inventory of previously adjudicated cases, the decision should have come as no surprise to anyone conversant with the history of Canadian jurisprudence. Here is all the more reason for arguing that the King ministry, using the expertise of its legal civil servants, had to have come to the conclusion that its case in favour of section 5 (a) being *intra vires* of parliament, would not stand up in court. Therefore a reference to the courts, thanks to Euler's motion in the Senate, was seen as a way out of a dilemma which involved political self-preservation.

The gist of the Court's position was that section 5 (a) had been legal when originally passed. Control of margarine had been perceived and accepted as a measure for the protection of public health, an area of federal jurisdiction. Since 1886, the manufacturing process had been so improved that public health was no longer in danger. In 1948, control of margarine was determined to be a matter of producing for sale a legitimate product, a property and civil right, within the jurisdiction of the provinces. Such a shift in perception would most likely have occurred even if the issue had come before the Supreme Court in another era – an era not marked by a heavy emphasis on provincial rights.

Reaction to the decision was in line with the original position of the litigants. Senator Euler probably felt even more satisfied, more jubilant, than when he won his first federal election in 1917. The urban daily press was unanimous in trumpeting margarine's victory with page-one headlines, and several expressing approval in editorials and cartoons. Very few raised the point that the Court had expressed only an opinion, that section 5 (a) would remain legally valid until parliament repealed it or that an appeal to the JCPC was still an option. The *Calgary Herald* set a somewhat hyperbolic tone:

> Other decisions of the Supreme Court have no doubt been more important, but few have been so anxiously awaited by the general public. And few have been made at a more auspicious time, for within the past few days butter virtually disappeared from the stores. Margarine just had to come.[8]

Public reaction was almost unanimously in favour of the Court's actions, if the surveys taken by several newspapers among their readership were any indication.[9] The needs of all Canadians as consumers had to take precedence over the needs of the producers of butter.

For the winners—cheers: for the losers—fears. The reaction of the latter ranged from the quite honest but extreme one of "betrayal" by the Ontario Cream Producers Association,[10] to one using similar terminology but with tongue in cheek. Bandleader Peter Parry had recently written a song entitled "Margie, Margarine," appealing to the government to lift the ban. The Supreme Court opinion had been handed down just before Parry and popular singing star Joyce Hahn could record it, wrecking Parry's chance of earning much in the way of royalties.[11]

The initial response by the National Dairy Council was "wait and see." H.H. Hannam, of the Canadian Federation of Agriculture, believed the situation had been left in disarray, in that jurisdiction was now divided—the provinces with control over manufacture and sale, the federal government with control over importation and public health standards. The Federation was considering an appeal to the JCPC, but was waiting for the federal government to make its decision on this step first. Paul Martin, the Minister of National Health and Welfare, was reported as not being prepared to state what the federal government's policy on the appeal would be, but that his department was ready to set proper health standards once it had been determined that the ban had ended.[12]

J.H. Duplan of the National Dairy Council commented that, contrary to reports, the Council was not contemplating seeking an injunction against the sale of margarine. The Council had already come out in support of the temporary introduction of margarine to ease the butter shortage. It was waiting for the federal government to clarify its intentions before reacting.[13] The Canadian Federation of Agriculture took the position that it was preferable for the federal government to carry through the appeal. The cost could have been borne by the Federation, but obviously it would be easier to have the federal government assume these charges. Moreover, the Federation's leaders feared a weakening of its negotiating position if it got sidetracked from its efforts to have Ottawa institute an alternative program—the removal of the price ceiling on butter and an expansion of the export markets.[14]

After some soul-searching the King ministry declined to appeal. The public argument given was that the government was intending to end all appeals beyond the Supreme Court: to make one further appeal almost simultaneously would not be very logical. The government had achieved its political end, shifting responsibility for a decision by reference to the Supreme Court: further action was unnecessary. Even more importantly, it would be politically impossible to maintain the margarine ban now that the Court had ruled it dead. Even if the JCPC reversed the Supreme Court's ruling, the federal government would be forced by public opinion to repeal section 5 (a). The government did go so far as to agree not to act to abolish appeals before the provinces or any private bodies had decided whether or not to appeal.[15] The logic behind such postponement was weak because it was also understood that any repeal legislation would stipulate that cases already before the courts could continue for a final ruling all the way to the JCPC. Thus the margarine issue could be appealed under any circumstance.

The failure of either level of government to take the initiative caused the Canadian Federation of Agriculture to act in March, 1949. This was done with some confidence because the dissenting opinion of Chief Justice Rinfret was understood as giving them a solid base upon which to argue before the JCPC in London.[16] The federal government at least was willing to support the Federation's action.[17]

Who bore the costs of the submissions to the two judicial bodies? In the nineteenth century, the proponents of butter had already raised the argument that the fight was between the Canadian farmer, a small native businessman and the giant national and even international business complexes. During World War I it had been, of course, the food corporations which manufactured margarine. In the summer of 1949, more concrete evidence demonstrated the existence of a connection between the various elements on the margarine side of the battle. In July, Euler wrote to Salter Hayden, his legal counsel, noting the approval by Lever Bros. of Hayden's move to oppose the appeal to the JCPC. Lever Bros. assumed the full financial responsibilities of fighting the Euler side of the issue.[18] On the other side of the case the federal government and the Dairy Farmers of Canada shared the costs ($4,989.00) of the appeal to London.[19]

It took until June, 1950, for the JCPC to hear the submissions.[20] On October 16, its ruling was handed down; margarine users in Canada could rest easy, since their supplies of margarine would not be cut off. The *Margarine Reference* was the JCPC's last consideration of the trade and commerce power of the BNA Act and one of the last six judgements handed down on any Canadian case. The judgement, based on the principle of *stare decisis* with very little social data being considered, again offered no surprise to those who knew the law and the system.

The JCPC ruled that section 5 (a) was *ultra vires* to the extent that it dealt with the manufacture, sale, offer or possession for sale of butter substitutes. The question of importation had not even been raised. The JCPC's decisions are given in the form of advice to the monarch and therefore are given as a single opinion without dissenting judgements being recorded.

Their Lordships first of all came to the conclusion that earlier decisions had clearly established that the federal government did not have the power to regulate individual forms of trade and commerce, especially as confined to one province. In the debate as to whether section 91 (2) or section 92 (13) applied to margarine, their Lordships concluded that the main purpose of section 5 (a) was to protect the Canadian dairy industry. Therefore again the federal government had no power to act in this manner. As to the agricultural issue, the federal government had failed to demonstrate that the ban related directly to the agricultural operations of farmers. Therefore the federal government again had no jurisdiction. Lastly the situation did not allow for an emergency to be declared; therefore the "peace, order and good government" clause could not be invoked by federal authorities to override the normal distribution of powers as laid out in sections 91 and 92 of the BNA Act.[21]

The anticlimactic decision of the JCPC really changed nothing except to confirm that the situation would continue to develop as it had done for the last 22 months. The first margarine manufactured in Canada in a quarter of a century was produced in Vancouver;[22] it went on sale on Monday, December 21, 1948. The legal decision relieved the federal government of most of its political responsibilities in this matter and shifted them to the provincial level. The decades ahead would see the fight between butter and margarine erupt on several fronts: federal and provincial politics, economics, colour, and nutrition.[23]

— *Courtesy Provincial Archives of Alberta, Pollard Collection, P4127, 78.63/1157*

PART THREE
Margarine Prevails 1950-87

CHAPTER EIGHT

The Legislative Maze

The butter industry was about to become the most regulated sector of Canadian agriculture: the margarine industry suffered much less from governmental paternalism. Yet the context in which the national and provincial governments reacted to the new legal situation had changed. In the past, agriculture and dairying had been very important components of the national economic, social and political order. In the post-war era, agriculture and dairying were still significant elements to whose needs politicians had to give heed, but their importance was declining. Agriculture employed the third largest number of people within the national labour force; within agriculture dairying was third in the size of its labour force behind mixed and grain farming, and third in the spectrum of revenue from agriculture.[1]

Provincially, agriculture and dairying maintained important rankings. For example, in labour force, agriculture stood first in the three Prairie provinces and second in Quebec. Fiscally, Ontario and Quebec earned almost 75% of gross farm revenue produced by dairying in Canada. In the production of revenue from agriculture, dairying stood first in Newfoundland, Nova Scotia, New Brunswick, Quebec, and British Columbia.[2]

This ranking provides no real clue as to why each province's government reacted to the new situation as it did: Quebec and Prince Edward Island imposed a ban on margarine; Ontario demonstrated serious reluctance to take on regulation of any kind; and Alberta was one of the leaders in aiming for as little regulation as possible and that only from a public health point of view.

Following the Supreme Court's ruling in 1948 the federal government came under pressure from the whole spectrum of special interests: the butter lobby pleaded for legislation which might get as close to a ban as possible. The industry recognized that it was part of a "minority group in the social structure today and we must convince governments that what agriculture asks for will be good for all."[3] This effort showed best in *More Tiny Drops*, a pamphlet issued by the National Dairy Council. Dairying was the direct or indirect source of income for 1/6 of all Canadians, but its dollar volume was nine times that of the fruit-growing industry, ten times that of fishing, two and one-half times that of the gold mines, eight times that of the coal industry, and greater than the dollar output of each of the pulp and paper industry and of lumber.[4] Small numerically, but vital to national well-being, the industry was in need of some consideration. The past decade had seen increases in production and consumption of most milk products, but a significant decline in the number of producers.

From its perspective, the margarine industry also understood its best interest to be served by government action. Its agitation came in the form of overt support from consumers' groups as well as covert lobbying.

Mackenzie King's ministry had been bruised enough by the margarine controversy. His ministry, and that of his successor, Louis St. Laurent, were quite happy to give up prime responsibility.[5] The areas of possible action by the federal government included the import ban, food and drug standards, new taxation, and support of the price of butter. Ultimately, security for the butter industry was the chief political consideration.

The government's initial effort to protect butter from the margarine threat was the levying, in 1949, of an 8% sales tax. Under the provisions of the Excise Tax Act, the sales tax came into effect automatically once margarine was legally in the marketplace. This action was taken in spite of the facts that American farm lobbies were abandoning calls for taxes on margarine because they had not proved effective in curbing sales,[6] and that the cost of living was climbing steadily.[7] The action provided Senator Euler with cause for another crusade related to margarine: a repeal of the tax could lower margarine retail prices significantly.[8] (In 1953, relief for Newfoundland was achieved when, by order-in-council, the sales tax on margarine shipped from Canada to Newfoundland was to be refunded.[9])

The tax also came into existence despite the federal government's continuing price support program for butter under the Agriculture Prices Support Act of 1944. The third federal action came in 1951 with a severe tightening of import controls on butter, cheese, canned milk, and dry milk powder: no dumping at any price was to be permitted unless licensed by the Department of Industry, Trade and Commerce.[10] James Gardiner, Minister of Agriculture, argued that the support system was in place to protect the consumer.[11]

In the area of control legislation a very minor shot was fired in the winter of 1949, when George Cruickshank continued his fight against margarine by introducing a private member's bill whose intent was to protect the dairy farmer and the public from "deception": the ban would have been reimposed. Bill 22 got no farther than first reading.[12]

Not until late spring of 1951 did the St. Laurent government bring out new margarine control legislation. The Dairy Products Act was the result of consultations with the provinces and the dairy industry. It repealed parts I and II of the Dairy Industry Act, 1927, thus ending the effort to protect dairy products through control of production and sale, and replaced this method by controls in the area of interprovincial and international trade. It also provided the power to establish standards of quality which would become operative when the product became involved in interprovincial or international trade. Continuation of the import ban came in spite of support of worldwide efforts to liberalize trade, as shown in the acceptance of the Geneva Trade Agreements which denied the right to prohibit the importation of any product which was legal in each participating nation. Even more significantly, the new legislation's intent ran counter to section 121 of the BNA Act which stipulated free interprovincial trade. After another strong rearguard action by George Cruickshank, who pointed out that it would do little to protect the butter industry, the bill passed the Commons.[13] In the Senate, Euler used the occasion of the longest debate of the session to mount another attack against the government's policy. This time he focused on clause 6, paragraph (c) which gave the Cabinet the power to regulate and even prohibit interprovincial trade in a product. He vigorously condemned what he saw as a violation of one of the basic premises underlying Confederation, a free exchange of goods. He lost his fight in the Senate[14] even though, belatedly, powerful elements of the media such as the *Toronto Star*[15] gave him its full support. *Maclean's* editorialized: "Neither House has cause for pride in this shameful default of duty. No party, including those in opposition, can claim exemption from the charge of appalling cowardice."[16]

Under pressure, the government gave up its effort to control interprovincial trade in margarine and any other butter substitutes. The tradeoff was to get several provinces to end their embargoes on cattle trade which had been imposed after recent outbreaks of hoof and mouth disease.[17] Gardiner acted by assuming sponsorship of the private member's bill introduced by David Croll in the Commons and Euler in the Senate.[18]

In one sense the passage of the Dairy Products Act in 1951 marked the culmination of two years of intense, concurrent activity by the provincial governments. This action took place on two levels: attempts at a common position and separate provincial initiatives.

Five weeks after the Supreme Court handed down its ruling, the National Dairy Council took the industry's first step to seek clarification of the uncertainty it perceived had been generated by the decision. A letter was addressed to Prime Minister St. Laurent, copies of which were sent to each provincial premier. The Council requested an industry/government conference to review the constitutional position of all federal and provincial legislation relating to the dairy industry. It also asked for a review of the overall economic conditions of the milk industry with a view to expanding, with government support, domestic consumption and exports.[19]

Prodded by this industrial initiative, supportive political action came from Douglas Campbell, Premier of Manitoba.[20] St. Laurent's reply was positive but low keyed.[21] About the only result that came out of these manoeuvres was an Ottawa meeting, three days after St. Laurent's letter had been sent. Nine industry delegates met with representatives from British Columbia, Saskatchewan, Manitoba, New Brunswick, Nova Scotia, and Prince Edward Island; no spokesman for the federal government was present. The meeting achieved little except a statement suggesting a year's suspension of manufacture and sale of margarine, until the JCPC's ruling would be handed down. The alternative was legislation which would be so restrictive that it would fall just short of an absolute ban: licensing of manufacturers and retailers, no production of a yellow-coloured product, and strict packaging and pure food standards.[22] The legislative steps ultimately taken by the several provinces were based upon an appreciation of the situation which had developed at this conference.[23]

Newfoundland, having enacted no unnecessary restrictions in the first place, was the only province to avoid political contortions over margarine. The other provincial governments took up new regulatory responsibilities. By October 1950, when the JCPC gave its ruling, all had taken some initial step. By 1951, with the passage of the federal Dairy Products Act, they had settled into their respective political positions. The history of this activity can best be appreciated if the provinces are grouped according to the level of restrictions which they placed on margarine.

As is often the case, federal legislation required enabling legislation in each province in order to become effective. No evidence of immediate action has been found with regard to the 1886 ban. In the 1930s, two provinces had taken such action. In March 1935, British Columbia was the first province to put such enabling legislation into force.[24] In 1937, Prince Edward Island passed a Dairy Industry Act which through section 2 made the federal ban operative, but the law was not immediately promulgated.[25]

Following passage of the 1951 Act, however, Prince Edward Island and Quebec led the way in very strong attacks on margarine. In both provinces, agriculture stood high on the list in proportion of labour force employed, but, in gross farm revenues the two provinces stood at the opposite ends of the spectrum.[26] Prince Edward Island, near the bottom in dairy revenue and ranked eighth in proportion of total production of butter in Canada, responded first with prohibitory legislation. The political climate in the province was one in which the fortunes of any branch of agriculture were of great concern. The Charlottetown *Guardian* had been keeping up a low-key commentary on the efforts in Ottawa to end the margarine ban. Its editor saw quite clearly the role of the protective impulse — in this case for farmers and politicians:

> as a piece of protection for dairy farmers when they were more influential politically than now. Urban consumers were indifferent at the time, since butter was cheap and plentiful.
> Today... the economic effect would not be important. Politically the removal of the ban is practically impossible. No party dare risk the certain loss of all the farm vote.[27]

The editor also bemoaned the Court ruling, arguing strongly against the disregard of the principle of parliamentary democracy.[28] However, in spite of the fact that the Island dairy industry had just achieved its highest annual gross production,[29] the provincial government became the first to act against margarine. On January 13, the Dairy Industry Act, originally passed in 1937, was finally promulgated. It banned manufacture and sale of any butter substitute.[30] Once the legislature met, surety was piled upon surety by the passage of another Dairy Industry Act which dealt specifically with butter substitutes.[31] The new legislation went beyond the 1937 act by making possession *prima facie* evidence of intent to sell. The act did not end the Islanders' newly formed habit of smuggling margarine from the mainland for their personal use and for resale to relatives and friends. The Islanders continued to protect their own interests as best they could.[32]

In Quebec, the economic importance of the dairy industry and the continuing efforts of Premier Maurice Duplessis to advance the cause of provincial autonomy — and his personal power — provided the context for that province's initial response to the margarine challenge. In his first comment on the impact of the Supreme Court's ruling, Duplessis warned that it would be imprudent for anyone to begin the manufacture and sale of margarine before the provincial government had made some decision.[33] His personal inclination was toward severely restrictive legislation.[34] However, a strong element of Quebec public opinion favoured legalizing margarine. Women's organizations sent three telegrams deploring the fact that a government, standing strongly for provincial autonomy, might deny that same autonomy to the individual within the home.[35] The Gallup Poll established that 55% of Quebec voters wanted margarine.[36] At the same time, the butter industry was being severely affected. Butter sales were being curtailed by margarine's availability to the point that second-grade butter could hardly find buyers at the retail level.[37] On March 4, 1949, the government rammed through the legislature (by 78 to 2) an act banning manufacture, sale and possession of margarine.[38] This step, intended to protect the dairy industry, was precautionary since the act would come into force by order-in-council only if the JCPC's ruling went against butter.

The only semblance of common sense in the debate was demonstrated when Frank Hanley (Independent: Montreal-St. Anne) suggested that legislation be passed to enable the production of margarine in the province in quantity equal to whatever gap might exist between production and consumption of butter.[39] No one paid any attention.

The proposal to ban possession created an impossible situation for law enforcement officers. In Hull, they moved to set up a "marge squad" in an attempt to police the traffic between Hull and Ottawa. Every car, bus and streetcar crossing the river from Ottawa to Hull would have to be searched; every person was to be considered a potential smuggler.[40] The government had not lost all common sense, and two weeks later, the order-in-council putting the act into force omitted the section on possession.[41]

For the other seven provinces March/April 1949 was also a period of intense legislative activity concerning margarine. Ontario has continued to hang on to restrictions as to colour and signs in public eating places down to the present day. However, the Ontario government's initial action was very close to those of its sister provinces. When the Supreme Court ruling was handed down just before Christmas, 1948, T.L. Kennedy was premier in a caretaker role in the half year between George Drew's and Leslie Frost's leaderships. Consequently he had no normal power base from which to function. As continuing agriculture minister and as a farmer himself, he had strong sympathies with the dairy industry. His response to this new challenge was publicly to adopt a complete hands-off attitude[42] while privately leaning towards at least a colour ban.[43] Kennedy's ministry could not avoid some official reaction. His government had to respond to pressure from the margarine industry to set some sort of policy within which Lever Brothers could establish the feasibility of constructing a facility for the production of margarine which would cost in excess of one million dollars.[44] The Ontario government rejected a request from the New Brunswick ministry for the suspension of manufacture and sale of margarine.[45] Kennedy's true sympathies were balanced by the realization that Ontario would gain by allowing the local manufacture of margarine for distribution to a large part of the country. His government's position was shown in the legislation passed in late March, which adopted the basic terms proposed by the Ontario Federation of Agriculture in its brief of February 25: colour restricted to a very light yellow, and control of packaging and of display signs in public eating places. The OFA's purpose was protection from unrestricted competition from what it still considered an imitation of, rather than alternative to, butter.[46] The bill passed virtually unanimously despite the protest raised by 1500 Toronto consumers.[47] Three years later the Edible Oil Products Act was enacted, forbidding any misleading statements on packaging and in the advertising of margarine. The law also forbade any packaging and advertising connecting margarine to any dairy product except skim milk, or making use of pictures of "dairy scenes." An inspection program was also established.[48] A year later, the Act was amended to prohibit the blending of edible oil products with butter.[49] Nothing was done to get the butter industry into a sound competitive position.

New Brunswick and Nova Scotia completed the pattern of a negative response toward margarine in eastern Canada. Like the other Atlantic provinces, New Brunswick and Nova Scotia contributed only a very small proportion to national dairying revenues. Yet for each province dairying was highly significant within the total agricultural revenue structure. New Brunswick was an exporting province; Nova Scotia, on the other hand, needed help to meet its dairy needs.

The New Brunswick government quickly came under pressure from both anxious dairy farmers and urban workers.[50] A.C. Taylor, Minister of Agriculture, attempted to secure suspension of manufacture and sale of margarine in the five eastern provinces to be affected by the new circumstances. He received positive responses from only Quebec and Prince Edward Island. Failing unity of action, the government moved to pass a regulatory act, but it did not give any serious consideration to a total ban.[51]

The terms of the bill were basically the same as those being passed by most of the provinces regulating public eating house announcements and requiring a light shade of yellow. No yellow colouring material was to be included in the package. No preservatives other than common salt were to be used.[52] This latter requirement was to generate some difficulty in the industry as margarine manufacturers, centred in Toronto and producing for all of eastern Canada, also used benzoate of soda for preservation purposes. The following year an inspection service was established.[53]

In the sister province to the east the protective impulse was much stronger. Nova Scotia suffered from a sizeable shortfall in the production of fluid milk and of butter.[54] Despite this fact, the provincial Deputy Minister of Agriculture had delivered a gratuitous public attack against the margarine forces' work long before the Supreme Court ruling was handed down.[55] This was part of a debate being carried on in the province in which the *Halifax Chronicle-Herald* was a vociferous exponent of the margarine arguments. As well, the government's pro-butter position was condemned as "defeatist" by the opposition.[56] The government's response came early in April. Having rejected New Brunswick's suggestion for a ban covering most of eastern Canada, the Liberal government introduced Bill 132, nearly identical to the New Brunswick law.[57] The bill generated a rousing debate, primarily on the colour issue. The pro-butter supporters again argued the protectionist line.[58]

Editorially, the *Chronicle-Herald* raised two points that demonstrated the lack of soundness in the pro-butter protective stance. The editor noted that the requirement for public eating places to display a sign informing patrons that margarine was being served in that establishment was "unusual" legislation. He forecast that the sign would, ironically, become good publicity for margarine. Secondly, why should butter receive such favourable treatment? Why not have signs "oranges are served here" to protect the apple industry; or "corned beef and cabbage are served here" to protect the fisheries; or "cake is served here" to protect the bread industry?[59]

Common sense played a much stronger role in western Canada. In the prairie provinces, dairying was only third or fourth in contributing to gross revenues from agriculture, but in British Columbia it ranked first. In Manitoba, a feeling of bowing to the inevitable was the most negative reaction generated by the issue. Reorienting the dairy industry would be necessary, and it could and would be done.[60] In return, the industry asked Canadian dairy consumers to help the dairy farmer by supporting a continental free market in dairy products. Such an external market for surplus Canadian production would mean higher prices for all milk products in Canada.[61] The 1948 annual convention of the provincial Women's Institute passed a pro-margarine resolution by a substantial majority.[62] In the legislature, the earliest step was taken by a private MLA who was a dairy farmer. His resolution called only for regulation of the product; the most negative aspect was the clause which stipulated that margarine ought not to be coloured to resemble butter. The motion was accepted by a non-recorded vote, but during the debate no attempt was made to move the intent of the legislature's action in the direction of a ban.[63]

Two months later, the government introduced and the legislature accepted its response to the new situation. It followed the moderate trend set in eastern Canada. One new element was to require name and address of the manufacturer on the package.[64]

The history of margarine in Saskatchewan is a fine example of a point that T.C. Douglas, CCF Premier of the province, had been making on numerous occasions — that people living within the economic order, especially the urban and rural sectors, are interdependent. This interdependence required people and government to function in ways which would recognize this fact and respond for the benefit of the total community.

Within the CCF caucus, and in the legislature, some pressure developed for a provincial ban.[65] In the Assembly, the opposition Liberals adopted a pro-ban position. Douglas replied that the best solution was for the federal government to continue to provide a uniform legislative structure for the whole country by setting up regulations under the Food and Drug Act and by banning the importation of all edible oils.[66] That same day, the Saskatchewan Dairy Association presented its brief to the government, arguing that margarine and butter needed to be placed within the same regulatory structure to make the competition in the marketplace a fair one.[67]

The government took until the second last day of the session to act, moving with the hope of getting as uniform a provincial response as was feasible, in particular as to colour and use in public eating places.[68] Consistent with its socialist orientation, the government feared a monopolistic control of margarine production. It was feared that such industrial giants as Canada Packers and Swift Canadian would use their power to destroy the butter industry and then obtain the equalization of prices. In this instance the principle of consumer's right of choice had to give way to protection of income for the small farmers.[69] The CCF national party's goal — the production of margarine by a crown corporation — although an alternative which met Douglas' personal and political views, would have to await CCF political control in Ottawa.

The common sense that basically prevailed in socialist Saskatchewan had a strong initial impact in conservative, Social Credit Alberta. However, the issue developed into one of the "most controversial" debates within and without the legislature.[70] In November, 1947, the Alberta Dairymen's Association had already made clear its position to Premier Ernest Manning. Butter was the product of an integral part of the Alberta dairy industry, providing 40% of dairy income; 60% of all milk produced in the province was turned into butter; half of the butter was exported to butter-deficient provinces such as British Columbia and Ontario. By providing a balance for agriculture, dairying occupied a central position in the mixed farming cycle. The damage margarine would cause would affect the whole province.[71] However, when the crunch came, the Association did not go so far as to demand a renewal of the ban, but rather asked for reasonable controls regarding the use of preservatives, labelling and standards of quality and advertising. The Association also wanted margarine to be a non-butter colour to avoid confusion.[72] From the margarine industry came pressure to leave the market a free one.[73] From the middle of the spectrum came a letter from a retired butter and cheese maker who urged Manning to leave margarine alone. Even the average farmer would want the economic advantage of using the cheaper margarine.[74] The Calgary Social Credit Constituency Association urged the provincial government to take no action against margarine, but to let it "be given an opportunity to sell on its own merits."[75]

Having perceived no consensus, it was quite logical for the Social Credit government to follow its natural conservative inclination and opt for free play in the market. On March 14, David Ure, Minister of Agriculture, introduced Bill 117 in the Legislative Assembly. The only proposed restrictions beyond quality control centred on public eating places using margarine being required to inform patrons, and proper packaging and labelling. Nothing was said about colour. The bill became the centre of considerable debate and, as a result, some alterations were made by the government before it became law six weeks later.

This legislation generated a strong renewal of efforts by butter proponents, but these focused on colour restrictions — even to the point of urging the prohibition of the inclusion of colour packets for home use. The Alberta Dairymen's Association and the Alberta Federation of Agriculture presented their new brief to the legislature's agriculture committee.[76] A local branch of the Canadian Legion carried on the fight for margarine by requesting no discrimination against it.[77] On March 24, the government proposed a new provision forbidding the sale of margarine in any colour that would allow mistaking it for butter. This action generated opposition from Social Credit MLAs. F.C. Colborne (Social Credit, Calgary) argued that both producers and consumers ought to be equally protected by allowing margarine in the marketplace, although he did not explain how that principle would work in practice.[78] Before final passage further rural pressure resulted in restaurants being unable to colour margarine for use on their premises.[79] A minor change the next year deleted the requirement that the word "margarine" be the "most conspicuous" word on the packaging. This step was taken to bring the law in line with that of the other provinces allowing the product.[80]

British Columbia's reputation for colourful people and history is enhanced by the margarine story in that province. The protective impulse helped generate a good deal of agitation on both sides of the question in a province which imported 82% of its butter.[81] Even though the dairy industry was a relatively small element in the total economic picture, dairying was a very important portion of the agriculture sector.

The 1935 Dairy Industry Act had been promulgated immediately. Attorney-General G.S. Wismer of the Liberal-Conservative Coalition ministry used this fact for his overly cautious ruling that margarine was under a ban even after the Supreme Court's ruling had been issued. The B.C. dairy industry had exerted strong pressure on Wismer on that specific point.[82] His position was taken despite the fact that section 4 of the Act stipulated that restrictions might be removed at any time by provincial order-in-council, and the sound argument that the provincial enabling legislation surely had lapsed once the federal act upon which it was based was declared *ultra vires*.[83]

The most outlandish example of the butter lobbyists' activity was that of the Shuswap-Okanagan Dairy Industries Cooperative Association from the dairy specialty area in the B.C. interior. Using "high pressure tactics and promotion" of an undefined nature, the president and the general manager attempted to influence the government's choice of action. But as rambunctious as the tactics might have been, the position expressed in the brief presented to the cabinet was comparatively mild — no ban was even suggested. Margarine advertising was to avoid any connection with a dairy product and all ingredients were to be listed on the wrapping. "For humanitarian reasons" no butter substitute was to be used at any provincial institution. Margarine ought not to be allowed to imitate butter as to colour or appearance. The provincial federation of agriculture took the same basic position in its annual brief to cabinet.[84] The agitation generated an editorial blitz from the metropolitan press. The *Vancouver News-Herald* stated it best. "The lobbyists went home to the Okanagan wondering if they had not done more harm than good! They may have won some converts among agriculturists, but on the whole lost ground by further stirring up emotions in the general public."[85]

Even without this stout opposition, the margarine forces in the province would have put up a strong fight. Two women were in the forefront of this effort: "Penny Wise" (surely a pseudonym), and Tillie Rolston. Wise, a shopping guide columnist, extended her crusade against the butter lobby's position and the government's reluctance to move promptly in favour of margarine entering the marketplace. In a week and a half, 1500 women — one quarter of her initial goal — had responded to her call for letters to Wismer.[86] Wise was joined by the railway unions, the Canadian Legion, the Vancouver Junior Chamber of Commerce, and by the B.C. Council of Women which would even have gone so far as to have the cabinet set the price of margarine in order to have the poor guaranteed the benefit of margarine's availability.[87]

This activity was aimed at influencing the politicians. In the cabinet the Minister of Agriculture, Frank Putnam, fought a lone battle against margarine, but one effective enough to prevent decisive action. An open vote in the Legislative Assembly was being suggested.[88] Twenty-five of the 48 MLAs responded to a newspaper poll: 14 Liberal-Conservative Coalition and 5 CCF favoured ending the ban, as long as confusion of margarine with butter could be avoided.[89]

From the government's back benches Mrs. Tillie Rolston, Coalition MLA for Point Grey, a metropolitan Vancouver constituency, also led the fight. Calling for an end to the federal sales tax and for choosing a palatable colour for margarine, she was instrumental in forcing the cabinet to come to a decision.[90]

On March 23, Bills 78 and 81 were introduced by the Minister of Agriculture. Both were passed the next day. Bill 81 amended the 1935 provincial Dairy Industry Act by withdrawing the federal Dairy Industry Act from the scope of the provincial law. Thus the way was cleared for Bill 78 dealing with margarine. In effect, the butter lobby had won its case. The protective impulse determined that government ought to have the power to regulate standards of quality, to keep advertising honest, to make sure that public eating places inform their patrons whether margarine was being served, and to regulate margarine packaging. Colour was limited to an ivory shade.[91]

The passage of the Oleomargarine Act, with its colour restrictions, set the stage for a two-year fight by Tillie Rolston to have this restriction removed. Against her stood the new Minister of Agriculture, Frank Bowman, who wanted no colour at all for margarine.[92] In March 1951, Rolston moved, seconded by E.E. Winch (CCF, Burnaby), that because the price of butter was becoming prohibitive and the butter shortage getting worse, colour restrictions on margarine ought to be removed. After the butter forces used the standing committee on agriculture to block a positive recommendation to the Assembly, Rolston withdrew the resolution. Her second effort, on an amending act, was defeated by a 2 to 1 margin.[93]

The only parts of Canada still requiring a decision were the North West Territories and the Yukon. The question was not dealt with by the Yukon Council. The Territories Council opted for permitting any margarine to be sold, so long as it met the requirements of at least one province.[94]

The first round of legislative responses to margarine's presence by the provinces and by the federal government had demonstrated a sharp division. Two provinces, Prince Edward Island and Quebec, opted for total protection—a return of the ban. Seven provinces attempted to balance margarine and butter interests and to achieve some semblance of uniformity across the country. But uniformity was not achieved and at times the margarine industry was faced with the requirement of different packaging for up to eight different situations. The context was an increasingly urban nation working the consumption of margarine into its eating patterns. Government action was also to be affected by a changing perception of margarine from a substitute from whose competition butter had to be protected, to a distinct food in its own right, purchased for health or economic reasons. In the coming years Newfoundland continued its open policy. Prince Edward Island and Quebec shifted from a ban to relatively open positions. The federal government and Nova Scotia, New Brunswick, Manitoba, Saskatchewan, Alberta and British Columbia loosened their hold. Only in Ontario was punitive regulation to continue.

CHAPTER NINE

Government Involvement since 1950

The federal government's involvement since 1950 has related to four areas: quality control, importation ban, sales tax, and lastly, prevention of fraud. Quality controls come under the jurisdiction of the Department of Health and Welfare. Over the past thirty years some significant developments have occurred. The United Nations' Food and Agriculture Organization/World Health Organization established the Codex Alimentarius Commission to set international standards for various foods. In 1970, the commission finished its work on a "Recommended International Standard for Margarine." Using this standard as a guide, in 1977 the federal government, after provincial consultations, set up Canadian national standards which have been used by the provinces to set their own standards.[1] (See Appendix 8.) The importation of margarine is still banned (Tariff Act, Schedule C, Regulation 99204-1), and there seems to be no political pressure aimed at ending this ban.[2]

The greatest amount of agitation has occurred in the area of the sales tax. Once margarine became legal in Canada, the Excise Tax Act came into effect and thus the applicable sales tax of 8% at the manufacturer's level was automatically levied (though not in Newfoundland). Increases to 10% in 1951, 11% in 1959 and 12% in 1967[3] helped to intensify the lobbying efforts against the tax *per se*. Active opposition came from the margarine manufacturers through the Institute of Edible Oil Foods (IEOF), the Consumers' Association of Canada,[4] the Fisheries Council of Canada and the Rapeseed Association of Canada.[5] Senator Euler added his efforts from the floor of the Senate. In 1960, the Senate rejected his amendment proposing the abolition of the sales tax.[6] The arguments used were consistent over the twenty years until success was achieved. The tax was highly discriminatory, the only federal tax levied unequally across the nation: it applied in nine provinces but not in Newfoundland. Margarine was the only food taxed in such a manner. As well, the tax most affected the standard of living of the poor, the people in greatest need of relief from the high cost of living.[7] The federal butter subsidy of 29¢ per pound meant a yearly bounty of $115 million to Canadians already rich enough to afford butter. On the other hand, Canadians who purchased margarine because of its lower cost were required to pay a premium — the sales tax.[8] These submissions were made not only directly to the Minister of Finance[9] but also indirectly to the federal government through briefs to the Royal Commission on Taxation, 1963-66, and to the special joint parliamentary committee on consumer credit, 1964-67. The Carter Commission recommended that margarine, "butter's direct competitor and inexpensive substitute," receive the same sales tax treatment and that these two foods, like all others, ought to be tax-exempt.[10] The parliamentary committee, appreciating that the Carter Commission had recently delved into this area more thoroughly, simply accepted the submissions from consumers groups and the IEOF and agreed that margarine ought to be freed from its discriminatory tax burden.[11]

Margarine's opponents at first were strongly opposed to the removal of the tax. They believed it a vital weapon in their fight to protect their existence as butter producers. By 1971, however, the opposition ended; the conclusion was that the tax served no protective purpose. The American experience helped sway Canadian attitudes. This viewpoint was symptomatic of a slight thaw in the war between the two contenders and a recognition of the foods as alternatives, not competitors.[12]

In the area of fraud, there has been a low-key running battle by federal and provincial law enforcement agencies to keep butter free of such bogus ingredients as lard.[13] The most complete tabulation of violations that has become public knowledge was that instigated by Stanley Knowles (CCF, Winnipeg North Centre) on March 3, 1954. In response to his query, he was informed that since July 1, 1952 eleven prosecutions for adulterating butter by adding vegetable oils had been successful and two cases were pending in the courts. All cases had involved Quebec companies; amounts seized ranged between 300 pounds and 26,000 pounds. Fines ranged from $25.00 to $50.00.[14] Ontario colour restrictions have been the single most important factor in generating law enforcement problems. For example, between 1981 and 1983, several hundred thousand pounds of margarine, manufactured by Ontario and Quebec processors for sale outside Ontario and therefore coloured like butter, were repackaged and sold as butter in Ontario.[15] Thus the federal government's role remains based on consumer protection for purity and fraud prevention.

Provincially, Newfoundland has remained on a sensible protective path. The latest provincial regulations, issued in 1966 by the Minister of Health under the Food and Drug Act, require that margarine be distinctly labelled, that ingredients be controlled and that certain stipulated amounts of vitamins A and D be included.[16]

In the years since the ban was imposed Prince Edward Island shifted to the position taken in most other provinces. The first modification came in 1965, despite protest by the provincial agriculture federation. This organization argued that most consumers had already reached a higher standard of living than Island dairy producers and that the presence of margarine in the marketplace would hurt the dairy industry even more.[17] The issue was not important enough to generate editorial comment in the *Guardian*, the long-standing proponent of legalizing margarine. An estimated half million pounds of margarine were being brought into the province annually, so the 1965 legislative action merely legitimized what was actually going on.[18] The legislative procedure to repeal the 1951 act and to put in place the Margarine Act took up the month of March 1965.[19] The Act set standards of quality and required proper package markings. It also stipulated that margarine had to be either lighter or darker than butter.[20] This legislation stood until amended in 1974, when the colour restrictions were abolished, diet margarine added to the definition of margarine, and minor changes made to standards.[21] Margarine was now in the P.E.I. marketplace under normal public health regulations.

In Quebec, the move away from the ban was just as inevitable. The province was Canada's largest butter producer, with production exceeding consumption. However, the ban was initially defended by the Duplessis ministry as a necessary protection for the province's dairy industry. Before long, it was found that maintaining the ban was just as impossible as it had been in P.E.I. Margarine itself became quite readily, albeit illegally, available at the retail level. An imaginative twist also appeared: labels such as Dari-spread and Champlain were given to a product that was composed of approximately 60% vegetable oils.[22] The government tended to ignore these facts until pressured. It then went virtually in the opposite direction in enforcement. Regulations were announced in 1952 to the effect that persons aiding in the conviction of another person for violating the margarine law would receive a reward of half the fine imposed.[23] The government extended the ban on margarine the following year and tightened up matters by granting cabinet the power to designate as a dairy product substitute any item the ministers chose to call a substitute, no matter what its name. In the process there was also added the stipulation that butter manufacturers who kept on their premises any ingredients which might be used in the manufacture of margarine would be subject to the same penalties as persons manufacturing or selling margarine.[24] The most extreme stance was yet to come: in 1961, the government banned the sale of colouring material for vegetable oil spreads.[25]

In 1961, the ministry began to face the realities of the situation. Continued bootlegging, as well as the fact that there was no quality control over the products being sold, led to renewed demands for margarine's legalization. The new legislation permitted margarine but only in a colour lighter than that set aside for butter.[26]

The Dairy Products and Dairy Products Substitutes Act of 1969[27] saw a general regrouping of the legislation dealing with dairy products and their substitutes. No change was made regarding colour, but as a result of regulatory powers contained in the law, the use of orange colouring was permitted shortly thereafter.[28]

In April, 1970, the newly elected government of Robert Bourassa was more prepared than its predecessors to listen to the submissions of a variety of Quebec interest groups. Circumstances were such that Quebec was the only province with a butter surplus, and its extra production helped feed the rest of the country. Seven other provinces were maintaining colour restrictions and thus acting to protect the Quebec dairy farmer from an increase in the margarine share of the bread spread market. Yet *Le Conseil de L'Industrie Laitière du Québec* argued that consumer protection was not being achieved by the mere regulation of the colour of margarine. Therefore, consumers ought to be given their preference in the colour of margarine. It also argued that the consumer would be protected to a greater extent if margarine were to be classified according to the nature and quality of the oil used in manufacture, and that proper labelling be required to instruct the purchaser on this point which most significantly affected quality and price. The *Conseil* went on to express the wish that the Quebec dairy industry should have the right to expand its own opportunities to use the province's milk surplus in a profitable manner by the removal of the restriction which made it the only food product manufacturer not to be able to produce margarine. As well, the industry wanted the right to mix margarine, butter and some concentrated milk products in the hope of creating a new outlet for the large quantities of milk produced in the province.[29]

The *Conseil* was thereby joining several other groups which had been calling for margarine to be given its proper place in the Quebec marketplace. The Quebec Food Council set before the new ministry the argument that the government ought to act in this area in ways that would lead to the establishment in Quebec of a manufacturing capacity for products for which there was a significant provincial market. The Food Council had joined the Grocery Products Manufacturers of Canada in making such submissions for more than two years.[30]

These addresses to the government were instruments of pressure based on two circumstances. First, for over two decades the Quebec consumers had had increasing opportunities to buy butter-coloured margarines and spreads, even though unlawful. Estimates were that three-fourths of the 55 million pounds of margarine consumed annually were coloured like butter and therefore illegal.[31] Second, declines in the country's ability to produce sufficient butter supplies also generated pressure to permit substitutes without restrictions. Regulations covering public health standards issued August 4, 1971 did not grant freedom to use alternatives. It was to take another year for the inevitable to occur.[32] The interests of the consumer won over those of the producer.

The seven remaining provinces had taken middle-of-the-road paths as their initial responses to the regulation of margarine. Six were to follow the path leading to P.E.I. and Quebec's positions.

In Nova Scotia in 1961 C.H. Reardon (Liberal, Halifax West) introduced a private member's bill proposing that the colour restrictions be completely removed. In a town/country split the bill was defeated 13-27.[33] A year later, having recognized the political realities, Reardon made an attempt to ease the colour restrictions. His new bill proposed as legal any colour outside the range used by butter. This version was easily passed, only four negative votes being cast.³⁴

Nova Scotia led the way in permitting blending. In 1972, the provincial Dairy Commission, with the support of the Milk and Cream Producers' Association, proposed a revision of the Margarine Act to permit blends of butter fat and edible oils. The industry recognized this opportunity to expand the market for basic dairy products. The blended product was to sell for less than butter, but the dairy farmer was to receive a better price for his butter fat.³⁵ This forward movement was made possible by the fact that enough controls were now in place to prevent a recurrence of the earlier widespread practices of adulteration. The action was also a response to the rapidly increasing costs of production in the dairy industry and the increased use of sophisticated substitutes, particularly in Europe.³⁶

The effort initially attempted by Reardon finally came to fruition in 1974. Submissions by the Consumers' Association of Canada, the provincial Council of Women, the fish producers, and others were made to the Minister of Agriculture to test public opinion. The agriculture lobby organizations raised no objection and Nova Scotia became the eighth province to end all but public health restrictions.³⁷

New Brunswick was also to move to regulations typical of those in effect in most provinces. Failure to achieve inter-provincial agreement to impose a common ban led to the passage of control legislation in 1949. This situation set the stage for a very intense effort to liberalize the colour regulations by the pro-margarine elements, among them the city council of Saint John and several other urban centres, the Ladies Auxiliary of the Canadian Brotherhood of Railway Workers and the New Brunswick branch of the Consumers' Association of Canada. Their contention was that such an essential product ought to be made as attractive as possible to the consumer. Moreover, the consumer did not need the additional task of working colour into a bag of margarine.³⁸

In response, provincial dairy industry organizations such as the Cream Producers' Marketing Board and the Dairymen's Association fought back with an equally intense effort. From their perspective, an outright ban was preferable, but obviously not possible politically. In lieu of the ban, existing colour restrictions ought to be maintained to provide the industry with as much marketplace protection as possible. The dairy industry portrayed itself as providing the economic undergirding for a "superior" way of life for almost half a million Canadian farmers and their families. Clearly, agrarian values were still being promoted as a vital part of Canadian life, with the argument that the well-being of Canadians was somehow related to "nature's most perfect food: milk."³⁹

Given the Hugh John Flemming ministry's strong orientation towards the dairy industry, it was logical to expect the passage of legislation in 1954 which would curb overblown advertising on behalf of margarine.⁴⁰ However, the margarine forces continued to exert pressure and this action, along with the general trend in many of the other provinces, led the government to introduce legislation which would permit margarine to be coloured at the factory, not only in the near-white range as before, but also in the darker-than-butter yellow.[41] Three years later even that restriction was removed.[42] In 1973, diet margarine was recognized by defining it as margarine with less than 49% fat in contrast to regular margarine which was to contain at least 80% fat.[43] Four years later, the Margarine Act and the Imitation Dairy Act were jointly amended to permit the manufacture and sale of blends of butterfat and edible oils.[44]

The four western provinces experienced a similar move to freedom from any regulation not related to public health. In Manitoba, an extended battle was fought over the removal of the ban on factory-coloured margarine. In free votes, nine years in succession, the Legislative Assembly voted down the bills of various private members aimed at ending the ban.[45] These rejections came in spite of strong evidence that many Manitobans were in accord with this aim. At urban municipal elections held in 1952, 65,000 votes in favour of factory-coloured margarine had been cast. In a follow-up, the *Winnipeg Tribune* had surveyed stores at 45 points in rural constituencies, finding that because it was a cheap, wholesome food, margarine had almost half the market. One rural store stocked only margarine.[46]

The government of Premier Duff Roblin finally acceded, in October 1959, to the call for an independent board of inquiry into the economic and social consequences of the colouring of margarine for both consumers and producers. William J. Waines, Dean of Arts and Science at the University of Manitoba, served as commissioner. The opportunity for another marshaling of arguments was taken up by both sides and 37 submissions of various kinds were received.

Three months later, Dean Waines submitted his report. He accepted the argument that an increasing price differential, not colouring, was the key reason why margarine was replacing butter in Canadian homes. Economically, the edible oil industry, relatively new but growing rapidly, was now providing the farmers with only 11% of gross farm cash income. Forecasts were that the dairy industry could not keep up with the milk needs of a rapidly growing Canadian population, and edible oils would be needed to fill the gap. Thus, in the long run, the economic advantages of margarine's presence in the marketplace would accrue to both producer and consumer. On the other hand, social benefits would accrue primarily to consumers. The consumer would be free of the task of kneading in the colour if the family did not want to use uncoloured margarine. In conclusion, Waines recommended the ban on factory-coloured margarine be removed.[47]

Waines' report failed to influence the government sufficiently to end the ban.[48] The next year it moved one step by permitting factory-coloured margarine, but only in a darker-than-butter shade. As well, to make sure fraud could not occur, public eating establishments, if they served margarine, had to present it in a triangular form.[49] It took until 1974 for all colour restrictions to be removed.[50]

For Saskatchewan, the story was also one of helping margarine find a place alongside the dairy industry in the province's economic structure. The Saskatchewan Dairy Pool was suffering a decline in milk receipts as farmers shifted to other, more profitable, products. Pool officials, seeking to protect their operation, proposed using the Pool's idle equipment to produce margarine. The vegetable oils would be acquired from seeds raised by Saskatchewan farmers and pressed by the Saskatchewan Wheat Pool oil plant in Saskatoon. Margarine would be made available in the retail market through the stores of the Saskatchewan Federated Cooperatives.[51] The concept was another excellent example of the interdependency point often made by T.C. Douglas. The interdependence of Canadians in Saskatchewan would be served by farmer-produced raw materials being turned, by urban workers, into a food available to both groups. Profits at all stages would be returned to the small entrepreneurs and consumers of the province. Certainly the market was there, as ten million pounds of margarine were being imported. The step was taken despite the protest by the provincial creamery association which feared a further erosion of its opportunities to sell butter. The raising of edible oil seeds by Saskatchewan farmers was slow in developing. Therefore the Dairy Pool was required to purchase refined oils from one of the soap companies. The cost was such that competition with the other manufacturers of margarine was impossible, and the plant operated only until 1955.[52]

The whole question took a long time to resolve. Douglas strongly supported bringing margarine into the economy as an essential item. He was able to block objections to the discretionary use of margarine by provincial institutions. The provincial Department of Health was saving $40,000 to $50,000 per year, the equivalent of 12-14 nursing positions.[53] However, in the area of colouring restrictions, the views of I.C. Nollet, Minister of Agriculture, prevailed.[54] In 1965, factory-coloured margarine, in a darker-than-butter shade of yellow, was permitted by the Liberal government of Ross Thatcher. The change brought the province in line with its two neighbours. The New Democratic Party, now in opposition, took a Nollet-inspired line of argument.[55]

By 1972, a dramatic change in the economic structure of the province had occurred and, in response, margarine was freed of colour restrictions. In the previous decade, a 14-fold increase in the acreage planted in rapeseed had taken place — to a total of 2.8 million acres, 50% of the total Canadian acreage. Here was the agricultural base the Dairy Pool had lacked in 1959. J.R. Messer, Minister of Agriculture in the Alan Blakeney NDP ministry, presented a range of arguments concentrating on consumer protection for their rationale. The government needed to protect the consumer by ensuring that food was "nutritious, healthful, and honestly advertised or sold." Personal preferences for good products must not be restricted. The government also had to avoid discriminating against rapeseed producers who were adding an important factor of stability to total income. A rapeseed processing industry would also contribute to the provincial economy's diversification. He foresaw no damage to the dairy industry resulting from this latest step regarding margarine. A slight decline in butter consumption was forecast but that would only aid in helping to strengthen the provincial milk manufacturing industry. Such action would help to reverse the decline in the number of cows (47%), in cream production (17%), and in butter production (46%) which had taken place since 1964. Even a cheese factory was part of Messer's dreams for his province, the vision which Douglas had expressed more generally two decades earlier. The Liberal opposition raised no protest, demonstrating that a consumer-oriented shift had occurred in Saskatchewan.[56] In 1973, the blending of butter and margarine was rationalized as an additional measure of support for the dairy industry.[57]

In Alberta the government continued its gradual easing of regulations affecting margarine. In 1952, the last opportunity arose for the dairy industry to prevent its unrestricted presence in the Alberta marketplace. A.J. Leisemer (CCF, Calgary) raised the question of allowing the sale of factory-coloured margarine provided that such action would not seriously affect the dairy industry. David Ure, Minister of Agriculture, rejected such an action because it would have the negative impact of upsetting the surplus production balance which provided the dairy farmers with an opportunity to increase their income through exports, in particular of butter. Leisemer objected in vain to this retrogressive position.[58]

By 1964, when the next step was taken, many changes had occurred in several sectors of Alberta's life. Consequently, legislation was passed permitting factory-coloured margarine in shades either lighter or darker than butter. The opposition and some government backbenchers protested that the "small, mixed" farmer would suffer from this new situation. In rebuttal, Highways Minister Gordon Taylor argued that he had polled his constituents, who were 4-1 in favour of a more open market. Coloured margarine would be "free enterprise in action" and competition would not necessarily be bad for butter.[59]

By the 1970s, the Alberta dairy industry was in decline, helping to generate a provincial and a national butter shortage. The prime factor was comparative costs. For example, one dairyman had left his farm in central Alberta because he could not make a living from a herd of 40 cows. He found success on his new farm in the Fraser Valley of British Columbia where he enjoyed a "very good" standard of living from 22 cows on 14 acres. A chance to put his cows in pasture for 10 months of the year, and the availability of cheap alfalfa from Washington were two reasons why costs were so much lower.

Meanwhile, rapeseed was now being grown in Alberta and the crop was of such a size that an oil production plant was in operation. The result was that 90% of all margarine produced in Alberta was made from locally raised and crushed rapeseed.

These two factors were the chief causes for the last major legislative action in 1972 which ended all restrictions relating to the colour of margarine but still maintained distinctive packaging and required public eating places to inform their patrons that margarine was being used. These remain in effect to the present day. A variety of factors helped to influence the legislators' minds. Other Canadian provinces had experienced continued growth in the dairy sector of the economy, despite a lack of restrictions on factory-coloured margarine. Moreover, consumer preference was based not so much on colour, but on price. Colour did influence consumer choice to the extent that it could affect the appearance of baked goods. Many Albertans admitted they took advantage of any opportunity to bring home B.C. margarine for personal use. Were a surplus of milk to develop, a specialty cheese industry would be preferred to margarine restrictions. The whole issue could be summed up in one question: was the government to protect the small, family-farm dominated agriculture of the past, or the new large-scale production unit epitomized by the rapeseed farmer? The new won out over the old.[60]

In the west coast province the work of "Penny Wise" and Tillie Rolston bore further fruit in 1952. British Columbia became the first of the provinces, other than Newfoundland, to end all colour regulations for margarine, ending one of the longest consumer battles in the province's history. In the shadow of an impending election and the thought of how annoyed consumers might react, the MLAs, with no real debate, voted 28-6 in favour of fewer restrictions.[61] Consumer protection had won. The province took one final step in 1973 by passing legislation which ended the requirement to have public eating places and margarine wrappers for the retail trade refer to margarine as a "substitute for butter." The argument was that B.C. was the only province retaining such a stipulation, forcing the manufacturers to make packaging specially for B.C. Secondly, given the new frame of mind of the general public, the phrase was no longer perceived as a warning that butter was not being purchased, but rather as an advertisement for margarine.[62] Peter Rolston, Tillie Rolston's grandson, now himself an MLA, was happy to see the fruition of his family's twenty-five-year crusade.[63]

Governed by a Progressive Conservative party that would eventually establish a record 42 years in power, Ontario had initially taken a reasonably progressive position on margarine. However, in the judgement of Dalton Camp,[64] successive ministries, especially that of William Davis, gradually shifted from progress to reaction. Thus, despite a changing demography and economy, Ontario dealt with margarine in an increasingly regressive way. The Ontario dairy lobby, one of the most powerful in the country, continued to have the ear of a succession of ministers.

The annual meeting of the Ontario Federation of Agriculture, in January, 1953, was the occasion for airing the positions of various sectors of Ontario agriculture, and for a compromise. Those perceiving margarine as the dairymen's worst enemy lost their attempt to call for a ban. The majority were prepared to recognize that margarine had won a place in the market, but further intrusions of synthetic dairy products had to be prevented. Behind this manoeuvre was the recognition that the 8,000 soybean growers were sensitive on the subject of a ban on edible vegetable oils. They represented the innovative side of Ontario agriculture and would have preferred no restrictions, even on coloured margarine.[65]

The compromise appeared in legislative form in the amendments to the Edible Oil Products Act. The government of Leslie Frost gave the agriculturists what they wanted; margarine was specifically exempted from the edible oil products to be prohibited from manufacture.[66] This response had taken place despite the intense effort of the Institute of Edible Oil Foods, the recently organized lobbying instrument of the edible oils industry.[67] At the same time, Albert Wren (Liberal-Labour, Kenora) tried to end the colour ban on margarine. His bill was rejected on second reading by a voice vote that crossed party lines. Wren failed by one vote to force a recorded vote and thereby embarrass some members.[68]

The complexity of the ongoing agitation between the two contending forces can be seen in the steps taken by each party. Despite hints to the contrary, the government declined to retreat. The Ontario Creamery Association attempted to achieve the effect of a ban by requesting the federal government to raise the excise tax from 10¢ to 15¢ per pound. Such an increase would balance cost factors between butter and margarine.[69] The provincial Concentrated Milk Producers' Association attempted to have the Ontario government stop margarine from being sold by home delivery dairymen.[70] The situation once again demonstrates the inherent conflict of interest among the various sectors of the economy related to the production and sale of butter.

The only change that has occurred in Ontario's colour legislation has been the acceptance of lighter or darker-than-butter shadings for factory coloured margarine.[71] This modification came as a result of continuing pressures from the pro-margarine forces, such as the Edible Oils Institute. Even when joined, two years later, by the Ontario Margarine Committee, it could not end all colour restrictions. In 1965, the Committee generated a province-wide petition by 50,000 persons in futile support of Leonard Reilly's (PC, Eglinton) private member's bill.[72] The 1963 Act also required the packaging to specify the type of refined oils and the percentage of each kind of oil used in the manufacture of margarine. The action was in response to the wide range of non-vegetable oils used in making cheaper margarines. Since the early 1950s, vegetable oils had been used to make 94% of all margarine; by 1962, however, that had declined to 63%. The consumer deserved to be protected by being provided with the information necessary to make a proper choice.[73]

The impact of "hardening of the arteries" suffered by the Progressive Conservative ministries was intensified by an outdated electoral map and the fact that the butter industry was part of a major segment of the provincial economy. In 1960, on average, one rural vote equalled four urban votes. In 1971, 26,080 dairy farmers helped make up that electorate, as against only 7,038 soybean growers. The dairy sector brought in $264 million – 11% of farm cash receipts; soybeans contributed a mere $27.9 million. Ontario remained a net importer of both kinds of products, but edible oil imports would mean foreign imports, while dairy imports were from Quebec.[74] The one factor which might generate some change in the colour restrictions would be an increase of the trade in illegally coloured margarine, as it had done in Quebec and other provinces. This trade has caused government inspectors to make raids similar to those in the prohibition era.[75]

The latest move to end the colour restrictions arose on a new front. The colour standards in the Ontario Margarine Act were those set by the U.S. Bureau of Internal Revenue. In 1985, a provincial court judge ruled that since these were no longer in force in the United States, they could not be effective in Canada. This action opened the door for "illegally" coloured margarine to be imported from Quebec to the degree that about 15% of Ontario's consumption was of this type of margarine.[76] The problem was left in the hands of the Liberal David Peterson administration as the Davis/Miller PC ministries lost control of the Ontario government. The Peterson government, after spirited discussion in caucus, declined to tackle another contentious area, gave in to the importuning of the rural MPPs and the butter lobby, and amended the Oleomargarine Act so that proper colour standards were in place.[77]

The fussing in Ontario has centred on the colour question. However, one other restrictive clause has remained in force over the years: public eating places must inform their patrons if margarine is served. The Dairy Inspection Branch of the Ministry of Agriculture and Food enforces it on a check-the-complaints basis.[78]

Another shot in the conflict was fired in April 1987, when the Quebec Minister of Agriculture, Michel Pagé, announced that colour restrictions, typical of those in force in Ontario, were to be imposed. The intent was to curtail the illegal sale of Quebec-made margarine in Ontario and to replace at least some of it with Quebec butter. He also hoped to reverse the ratio of 4.1 kilos of butter to 6.4 kilos of margarine being consumed annually by Quebeckers and move closer to the 5.3/3.6 butter/margarine ratio existing in Ontario. "We are the dairy province of Canada. We protect other industries, such as automobiles, why not the farmers," was the substance of Pagé's rationale.[79] Pagé's remarks demonstrate the impact of the intensified militancy of the dairy industry in response to the increased agitation of the margarine industry. In war, one shot requires returned fire. The stage was set for a renewal of the war between the two food industries.[80] Repercussions are quite probable in other provinces.[81]

In the 1950s, regulatory action was taken on the basis of a perception of margarine as a butter substitute. Two decades later, margarine had come to be seen as a distinct food used for health or economic reasons. Three and a half decades, and the growing complexity of the food production and processing industries, have also been the setting for the allocation of responsibilities to each level of government. The continuing urbanization of Canadian society generated changes in government positions. Each of the two sets of vested interests attempted to force government to support its perceptions — dairying to protect its past dominance and margarine to open the door of opportunity. The comparative economic advantage, and the nutritional advances, would help margarine expand its share of the market. The colour issue would be used in such a way as to restrict the growth of margarine so long as margarine's colour could not match that of butter. The regulatory maze would also serve as the bureaucratic framework for the ongoing margarine/butter conflict.

CHAPTER TEN

Economics of the Butter and Margarine Industries

Over the past four decades the butter and the margarine industries have each played a vital role in the national economy. Each has helped to develop and protect a high standard of living. Butter production, in particular, no longer the disorganized activity of farmers somewhat haphazardly meeting demand, has been intensively consolidated under federal and provincial regulatory bodies. Margarine has continued to be the product of an industrial complex but has become much more dependent on domestically grown edible oils, in a sector of Canadian agriculture different from that of dairying. Meanwhile, government policy has aimed at self-sufficiency in dairying in general and in fats and oils in particular.

Despite the belief of past generations that dairying would be the economic base for instant national prosperity, there were limits to productivity; climate and soil were the prime governing factors. The national herd reached a peak in 1937 of 3.8 million cows, shrank to 2.1 million by 1971, and has since then remained relatively steady. Consistent efforts to improve herd quality ensured that the total amount of milk produced continued to expand for another twenty-seven years after the national herd size had reached its 1937 maximum. Milk yield was increased from 4,287 pounds in 1939 per cow to 5,363 pounds in 1955 and 8,008 in 1975, though still below average yields in Europe and the United States.[1] The approach has been multi-faceted. A 1979 forecast predicted that improved herd nutrition, extension of the use of artificial insemination, and a reduction in the incidence of mastitis would increase annual production per cow by up to 30%.[2]

In the mid-1950s, the Royal Commission on Canada's Economic Prospects gave the country an opportunity to chart its economic future. The declining purchasing power of butter (Table 10-1) demonstrated a need for greater efficiency. In its brief to the Commission the Canadian Federation of Agriculture pointed out that the forecast of a 54% increase in population by 1980 represented the biggest factor in future food demand. Feeding 24 million citizens would require a 78% increase in meat supplies, a 31% increase in vegetables and 31% increase in wheat. For the dairy industry, the challenge took the form of increasing production of butter by 38%, fluid milk 54%, cheese 73%, evaporated milk 80%, skimmed milk powder 85% and ice cream 155%.[3] A study of Canada's import needs forecast that by 1980 total fats and oils consumption would increase by 121% over what it had been between 1951 and 1955. Canada was in the last stages of a transformation from a significant exporter of foods such as dairy products to the point where, except for grains, it had become a significant importer of food.[4] Yet the CFA could only rather lamely recommend a program of price supports.[5]

Table 10-1
Purchasing Power (Wholesale) of Agricultural Products

	1925-30	1950-54
Butter	100	84
Wheat	100	79
Cheese	100	138
Beef	100	235

Source: W.M. Drummond and W. Mackenzie, *Progress and Prospects of Canadian Agriculture* (1957).

The margarine industry submitted to the Commission a much sounder appreciation of economic realities. Any effort to expand total food supply ought to be recognized as contributing to the national welfare. From the long-range perspective, even Canadian agriculture would be far ahead if the edible oils used to manufacture margarine were grown by Canadians.[6]

The inherent problem appears more clearly when one shifts to a per-capita analysis. The peak production of 1,436 pounds of milk per person had been reached in 1942, followed by a consistent decline to 752 pounds in 1975. Domestic milk consumption followed a parallel pattern, a peak of 1,244 pounds per person in 1942, and a decline to 747 pounds by 1975.[7] Expanding production and declining per-capita consumption were being offset by the needs of a vastly expanded total population. The dairy industry was barely able to keep up with national demand for milk in all its forms.

This pattern of peak and decline can also be perceived within the spectrum of uses to which milk was put. Butter maintained its position as the single largest user, although its share of raw milk supplies shrank significantly from 60% in 1942 (the year of highest consumption) to 36% in 1975. In the three and a half decades from 1940 to 1975, per-capita consumption of milk dropped by 43%, and that of butter by 58%; but consumption of cheese actually rose by an astounding 383% and of ice cream by 327%. Butter consumption has continued its decline to the present day.[8]

The financial picture also provides some useful insights. Dairying consistently held third place behind field crops and livestock in generating cash income.[9] The dairy farmers had struggled through the Depression and had gone all out to meet the food needs generated by World War II. They entered peacetime anxious to share in the good times being promised by the politicians. "Doing well but there are problems" correctly expressed their economic circumstances. In 1947, the Bank of Canada calculated the profitability of farming in Canada. Using the average prices received by farmers between 1935 and 1939 as the index base of 100, farm income had moved from 88 points in 1935 to 183.5 in 1946. Costs of running a farm were calculated to have risen from an index point of 101.1 in 1938 to 149, while the personal cost of living had climbed to 130.5. The bank reached the obvious conclusion that farmers were ahead of Depression-year circumstances.[10]

The bank's calculations covered farmers in general, but dairy farmers were sharing in that greater wealth. The 1947 index of wholesale prices of selected farm products showed milk and its products in second place (185.2) behind livestock (204.7) and ahead of fresh meat (185.1), grain (167.7) and eggs (152.4). By 1960, the milk index had risen to 246.3, but milk had dropped to third place behind fresh meat (300.4) and livestock (292.4).[11]

Butter occupied a very particular place in the dairy sector of Canadian agriculture. Most butter came from milk cows grazed on land not suitable for the intensive farming that produced wheat, soybean or rapeseed. Nor was that land close to urban areas, in which case the milk would have been utilized as fluid milk. Milk produced in hinterland regions was processed into butter and other dairy products which could then be transported to areas of consumption.

Most Canadian butter was made in the half-year between May and October, the peak months for good pasture and for the greatest quantities of milk. Because butter keeps reasonably well when frozen, butter production using the peak months' excess milk increased dairying profitability. Skim milk, the by-product of cream for butter, gave the farmer further opportunity to supplement his income by raising calves, hogs and poultry.

Although butter production found a sound niche there were problems in the postwar years. Mechanization had benefited other aspects of agriculture, but the butter industry was still labour-intensive. Low wages and profits did nothing to attract young people as employees or owners. Butter production stood at the low end of several studies on dairying income and cost. There were much higher returns in raising beef cattle or even in selling milk cows to American dairymen.

Federal intervention through the Agricultural Price Support Act (1944) and the Agriculture Stabilization Act (1958) was intended to generate a constant demand, but at a consistently low price by maintaining floor prices. In the mid-1950s the price support system helped generate a wholesale price for butter of about 60¢ per pound. New Zealand butter was being delivered to Britain for a wholesale price of 45.6¢ per pound. Thus price subsidy and a prohibition on private importation of butter created a situation where Canadian consumers were supplying extra income annually to the butter industry—about 5¢ per pound because of the domestic price support and 9¢ per pound because of the import ban—to a total value in 1954-1955 of $44.1

Table 10-2
Indexes, Creamery Butter Production and Farm and Retail Prices
(1949-58 monthly averages)

Month	(Annual Average = 100) 1949-58 Monthly Average		
	Production	Farm Price	Retail Price
January	41.9	103.1	102.6
February	36.4	103.1	103.1
March	52.0	104.5	103.5
April	85.1	99.7	102.6
May	132.5	97.1	97.7
June	178.7	96.6	96.9
July	166.4	96.9	96.8
August	151.8	97.6	97.1
September	131.9	98.6	98.2
October	105.8	99.8	99.1
November	66.4	101.4	100.2
December	51.1	101.8	101.5

Source: *Report of Royal Commission on Price Spreads of Food Products* (1960).

Table 10-3
Profitability of Various Uses of Raw Milk

	Cost[a] Feb. 1947	Income[b] Oct. 1948	Income[c] 1949
	(per hundredweight of milk)		
Fluid Milk	$3.65	$4.02	$3.16
Concentrated Milk	3.00	2.97	2.39
Ice Cream	not given	3.17	2.28
Fluid Cream	not given	2.95	not given
Cheese	2.88	2.70	2.20
Butter	2.54	2.55	1.94

a. "Reviews Problems of Dairy Industry," *Guardian*, 20 February 1947, 5.
b. "Why Imports Are Necessary," *Moncton Daily Times*, 22 October 1948, 4.
c. *Report of Royal Commission on Prices* (Ottawa, 1949), 45.

million. There was a similar subsidy for margarine, since American margarine was selling for almost 8¢ per pound less than the Canadian product: 4-5¢ per pound credited to the import ban and the rest for the manufacturing cost differential. The total support figure came to $8.9 million annually.[12]

A fixed floor price made no allowance for rising costs of production: industrial milk producers faced a growing cost/price squeeze which forced them to accept a steady decline in real income since the beginning of the war. In 1962, to stimulate demand for dairy products such as butter, a direct-to-manufacturer 12¢ per pound subsidy was paid, thus bringing the retail price of butter closer to that of margarine. That same year, the Agricultural Stabilization Board also raised the floor price on both butter and cheese from 80% of the ten-year average to slightly above that average. This intervention stabilized prices at a level too low for profitable, full-time dairy farms, but consistent enough to attract and keep in operation a large number of farms with 12 or fewer cows, combining dairying with other types of farm operation in order to provide a more stable income than that from grain, cattle, or hogs alone.

After nine years of a growing cost/price squeeze, militant farmer protest in Ontario and Quebec forced the federal government, in 1967, to act radically through the new Canadian Dairy Commission (CDC), established in 1966. Beginning in 1965, the Pearson government had paid subsidies directly to industrial milk producers. To counter the impact of this policy, which encouraged marginal producers, the CDC used its subsidy eligibility quota to place minimum and maximum limits on the production eligible for price support. As a result of this stimulation, plus a phasing-out payment, over 70% of the small producers pulled out by 1971.

Table 10-4
Farms Reporting Milk Cows, 1961-71

No. Cows	1961	1966	1971	Percentage Change 1961-71
1-2	65,350	47,240	34,000	−48
3-12	160,500	95,000	46,300	−70
13-32	74,400	64,100	45,500	−38
33-92	8,500	13,700	18,500	+117
over 92	192	330	567	+195
Total	308,980	221,850	145,318	−53

Source: Don Mitchell, *Politics of Food* (Toronto: James Lorimer, 1975), 124.

This action eliminated those farm operators who were not able to keep up with the efficient, mechanized agriculture being developed in the 1970s. The 1966 census had shown that 54% of the industrial milk shippers had no electric milk equipment; 37% had no refrigerated storage; and 64% earned over half of their income from sources other than milk sales. This assault on the small shippers resulted in a 20% production drop in six years and led to a shortfall in industrial milk supplies. Increases in prices to the farmer from $6.17 per hundredweight in 1973 to $9.41 per hundredweight in 1974, and $10.02 per hundredweight in 1975 did little to offset the shortfall.[13]

Over the past ten years the supply of industrial milk has not met the needs of the nation as a whole. This situation has forced the adoption of market-sharing quotas and other devices to divide available supplies. In Ontario, in particular, the butter industry may not be able to continue as a viable economic entity.

The high degree of government intervention has both strength and weaknesses. Stability of prices — albeit depressed — and security of market values have allowed producers to plan their operations. The weakness is that there is no guarantee of net income since there is no protection against the cost/price squeeze for the producer. The second problem is that profit-taking by corporate agribusiness is not controlled. This has led to large-scale dairy processing being one of the most profitable businesses in Canada.[14]

The reappearance of margarine in the marketplace was not the sole cause of the steady decline in butter consumption. A proper picture of Canadian consumption patterns needs to take the whole range of edible fats into account. In 1953, while almost all consumers (90%) kept butter on hand and 66% kept lard and shortening, only 55% kept margarine. Slightly more than half the margarine was used in cooking, where it competed with butter, lard and shortening. For pies and pastry, lard and shortening were much more popular than butter and margarine.[15] Consumption of edible fats increased by about 50% between 1960 and 1982.

Even government policy generated another hurdle. During World War II the United Kingdom had been in much more urgent need of cheese than of butter and the Canadian Wartime Prices and Trade Board had stipulated a much higher price for milk shipped to cheese factories than to butter factories. This policy, as well as the pressures of consumer demand, generated the following picture of the price paid per hundredweight for milk to be used for different purposes: fluid milk $3.16, concentrated milk $2.39, ice cream $2.28, cheese $2.20, butter $1.94. With the end of wartime government involvement through subsidies and price controls in May-June 1947, the scales tipped back in favour of butter. Government policy from 1947 onward aimed at self-sufficiency.[16] More recently, and during the last two decades in particular, a very rigid marketing system under the direction of the Canadian Dairy Commission has regulated domestic supplies on the basic principle of supply management, with quotas for each province.[17] Butter continues at the bottom of the priority list as far as allocation of raw milk for various uses is concerned[18] in spite of consistently being the product line in which more milk is used.[19]

While per-capita consumption of butter has dropped almost steadily from a peak of 32.3 pounds in 1942 to 11.1 pounds in 1985, the pattern of margarine per-capita consumption has risen from an all-time low of 5.5 pounds in 1949, the first year of its legal existence, to a peak of 10 pounds in 1982. Since 1975, there has been a greater per-capita usage of margarine than butter.[20]

The key factor leading to margarine's inroads into butter's domain was, initially, the price differential. Colour and nutritional value came into play as manufacturers offered an ever-wider range of margarines in the following decades.

Butter's wholesale price underwent a consistent rise from a Depression low of 22.6¢ per pound in 1932 to a high of 83¢ per pound in 1975.[21] The steady price increase of butter can be seen in Graph 10-1. Initially, margarine was priced at 34.7¢ per pound retail, at a time when butter sold for 60.3¢ per pound. The price differential of nearly half has held quite closely since 1950, as can be seen in Graph 10-2. This picture is typical of that found elsewhere in the industrial world.[22]

Graph 10-1
Average Retail Prices and Price Relatives for Selected Time Periods, Canada

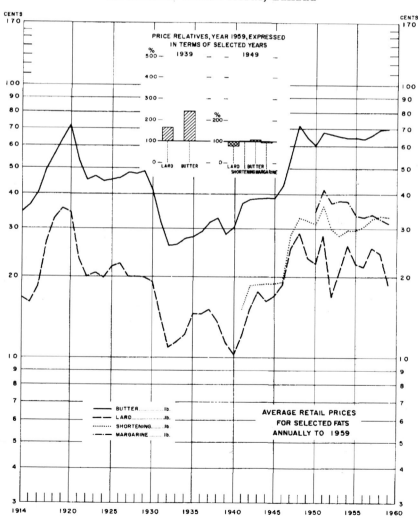

Source: Dominion Bureau of Statistics, *Urban Retail Food Prices, 1914-59* (Ottawa, 1960), 61.

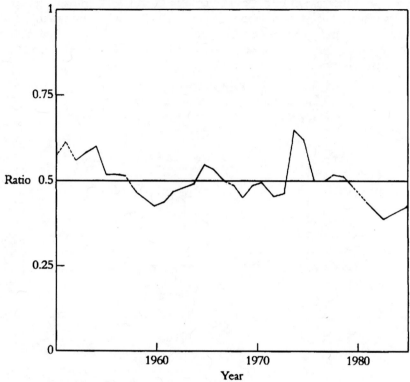

**Graph 10-2
Margarine/Butter Price Comparison, 1950-85**

Source: Statistics Canada, Can. Sim University Base (Ottawa, 1986).

During the transition from war to peace (1945-49) Canadians began to insist on the availability of margarine. That demand was stimulated by a 14.5-point increase in the cost of living, in contrast to only a 6.8-point rise for the six years of World War II. Since 1949 (to 1975) there was an 80.5 point increase in the index. With specific regard to food, the 1945-49 years saw an even more dramatic rise of 21.6 index points, followed by a slower but sustained rise between 1949 and 1975 of 100.6 points. The price of butter made a real contribution to the escalating cost of living. Between 1945 and 1949 the wholesale price rose 25.1¢ per pound (from 36.9 to 62.0); in the years 1949 to 1975, the price rose a further 21¢ per pound (62.0 to 83.0).

Another illustration of the impact of cost on the decision of Canadians to buy either butter or margarine is found in the Agriculture Canada studies of per-capita weekly food purchases by family income quintile group for 1969 and for 1974. Between these years the Consumer Price Index, for food alone, had gone up 46.7 points.[23] The per-capita weekly food purchases for all Canadian families for margarine had increased from 0.059 to 0.090 cents; for butter it had declined from 0.118 to 0.086 cents. In both years, there was an almost consistent pattern: the wealthier the family, the less margarine and the more butter was purchased.

Table 10-5
Per-Capita Weekly Food Purchases by Family Income Quintile Group, Canada, 1969 and 1974

Quintile Group	1969 Kg/Week		1974 Kg/Week	
	Margarine	Butter	Margarine	Butter
All	0.050¢	0.118¢	0.080¢	0.086¢
1st	0.070	0.110	0.104	0.096
2nd	0.079	0.112	0.082	0.092
3rd	0.060	0.122	0.079	0.032
4th	0.054	0.128	0.084	0.086
5th	0.045	0.121	0.069	0.086

Source: Agriculture Canada, *The Apparent Nutritive Value of Food Available for Consumption in Canada, 1960-1975* (Ottawa, 1981), 114, 116.

Cost-conscious Canadians, anxious to achieve and protect a rising standard of living, looked to margarine for relief. This conclusion echoes that reached by the Waines Commission in Manitoba in 1950. The Commission report also drew attention to American evidence that Canadian experience paralleled that of the Americans.

A second and, in the long run, equally influential factor in margarine's advantage in the marketplace was the technological revolution which provided the consumer with a vast choice in cost, quality and type. Margarine is an excellent example of what can be done in the area of food technology.

One additional example of this adaptability is the success the Canadian armed forces had in developing a biscuit spread for field rations. The product was required to remain stable without refrigeration for up to two years, to be produced from fats of Canadian oils, and to maintain its spreadability over the wide range of temperatures that would be met in Canada. A product composed of tallow and soybean oil met these needs.[24]

The wide range of edible oils and fats available from Canadian agriculture provides an almost unlimited base for continuing technological innovations. All margarine oils are interchangeable. There is nothing "infuriatingly unique" about any one of them, as there is with butter-fat.[25]

The Supreme Court's 1948 ruling which legalized margarine stimulated Canadian efforts to adapt various oil seeds to cultivation in Canadian conditions. The work of Prairie Vegetable Oils Ltd. to supply rapeseed oils to Britain during World War II was an example of earlier Canadian activity in the field.[26] In the early 1950s, there was some effort in the fishing industry to expand the use of seal oil to manufacture margarine.[27] However, it was

Table 10-6
Major Margarine Developments, 1890-1967

Year	Company (Brand)	Development	Significance
1890	U.S. manufacturers	Cultured milk and salt replace casein and salt as flavour source	Improved flavour
1917	Best Foods (Nucoa)	Vegetable (coconut) oil replaced animal fat	Better eating texture than animal-fat margarines
1923	Best Foods (Nucoa)	First to add vitamin A	Until this breakthrough only butter supplied vitamin A
1934	Best Foods (Nucoa)	Hydrogenated domestic vegetable oil replaced coconut oil	Desirably less firm, when cold, than butter and more nutritious, with P/S ratio of 0.20
1945	Best Foods (Nucoa)	Flavour protectors added	Enhanced flavour stability
1947	Best Foods (Nucoa)	Carotene, form of vitamin A, replaced coal-tar food colourings for margarine	Precoloured margarine available in certain states with a tax imposed
1950	Standard Brands (Blue Bonnet)	First to package in foil wrappers	Improvement over packaging of butter
1952	Kraft Foods (Parkay)	Blend of hydrogenated vegetable oils	Advertising theme: "Spreadable even when ice cold." P/S ratio of 0.40
1952	Best Foods (Whipped Nucoa)	First to produce soft whipped margarine in tub form	Ahead of its time; several years elapsed before consumer would pay premium price for added convenience
1955	Best Foods (All margarines)	Skim milk and flavouring used as flavour source; preservatives in packaging	Protection against microbiological spoilage
1956	Lever Bros. (Imperial)	Hydrogenated vegetable oils with small percent butterfat added	For the first time margarine required refrigeration and consumer was ready to pay more for assurance of freshness; P/S ratio of 0.40
1957	Kraft Foods (Whipped Miracle)	Commercial production of soft whipped margarine in stick form	Six sticks of the same weight for the price of four

Table 10.6 — *Continued*

1958 Standard Brands (Fleischmann)	Hydrogenated corn oil	P/S ratio of 0.40, same as that of conventional margarines
1958 Pitman-Moore (Emdee)	Liquid corn oil and hydrogenated vegetable oil (coconut oil)	Drugstore shortening-like product packed in tin can but with high nutrition — P/S ratio of 1.50, emphasizing polyunsaturated fatty acids but relatively high in saturated fatty acids
1959 Best Foods (Mazola)	Liquid corn oil and hydrogenated vegetable oils	First grocery-store margarine with very high (28%) polyunsaturated fatty-acids content, low (18%) saturated fatty acids content, and 1.55 P/S ratio
1962 Best Foods (Unsalted Mazola)	Skim milk as flavour source and protectors for added flavourings	Exceedingly long shelf life for this type of product and stability against microbial spoilage for consumers who prefer salt-free spread
1965 Best Foods (No-Burn Mazola)	Flavouring in aqueous phase and protectors	For smoke-free frying and no charring in cooking; excellent flavour release
1966 Best Foods (Mazola and Nucoa)	Eliminated pro-oxidants from salt in margarines	Enhanced flavour stability
1966 Best Foods (Diet Mazola)	Liquid corn oil and hydrogenated corn oil	Fat content cut in half as compared with butter and conventional margarines; improved nutrition; polyunsaturated fatty acid content, 44%; saturated fatty-acid content, 17%; and P/S ratio, 2.6
1967 Swift & Co. (Sundrop)	Pourable, nonseparating margarine	Emulsion remains uniform and pourable under refrigeration and room-temperature conditions

Source: Daniel Melnick in *Journal of Home Economics*, 60, 10 (December 1968), 796.

in the area of vegetable oil seeds that the search for raw materials for margarine was focused.

The search for the most useful source of vegetable oil concentrated on rapeseed. Its disadvantages were a high (10%) content of linoleic acid and an unacceptable (50%) level of erucic acid. The first created problems of flavour reversion, the second affected the balance between unsaturation or stability and requisite melting-point consistency.[28] A third problem affecting efficiency of use (and consequently farmer profits) lay in a low protein and high fibre content in comparison to soybean, thus limiting its use as a livestock feed. One great advantage was a 40% oil yield when crushed, compared to 17% for soybeans.[29]

The work of R.K. Downey in Saskatchewan, Frithjol Hougen and Baldur Stefansson in Manitoba, Sonza Machado and Wally Beversdorf in Ontario, and other persons at the National Research Council, contributed to bringing rapeseed — or canola as its genetically restructured strain is called — to the forefront of the Canadian agricultural economy.[30] By 1985 it ranked third behind wheat and barley as a money-earner among Canadian field crops.[31]

Table 10-7 provides a picture for selected years of the increase in the number of acres growing the three chief domestic oil seeds. Canadian production of these oils lessened dependency on importation of oil products, thus undercutting one of the major complaints raised by the butter forces. By 1963, the industry had made Canada the tenth largest producer of margarine in the world, generating dreams of competing in the world export market.[32] These expectations have not born fruit,[33] and the industry has centred on meeting only the needs of the domestic market.

Table 10-7
Land Seeded to Domestic Oil Seeds

Year	Canola (rapeseed)	Soybeans	Sunflower
1941	—	11,000 acres	n/a
1943	3,000 acres	36,000 "	n/a
1948	80,000 "	94,000 "	n/a
1951	7,000 "	155,000 "	22,000 acres
1961	710,000 "	212,000 "	34,000 "
1971	5,306,000 "	367,000 "	239,000 "
1981	1,492,000 ha	279,000 ha	116,000 ha

Source: *Historical Statistics* (1983), series M267; *Grains and Oilseeds Review* (December 1981), 10; and *Canada Yearbook* for pertinent years.

The impact of growing these edible oils on the agriculture and food industries was not only an explosion in production but also, and primarily, greater economic efficiency, providing the consumer with a cheaper product. Butter has lost ground in all comparative studies.[34]

To counter margarine's competitive edge in cost and versatility, the butter industry concentrated on maintaining its position as the "Cadillac" of bread spreads, when comparing texture, taste and colour. The dairy industry made efforts to develop its own butter substitute. Dairy interests in Montreal developed a product which consisted of butter with added milk and gelatine. It had 50% butter fat rather than the 81% required of butter, and was to have cost 15¢ less per pound.[35] It appears that this product never reached the store shelves. In 1951, the Ontario government made the facilities of the Ontario Research Foundation available to help the dairy industry create a dairy spread,[36] but no results have been recorded. The armed forces also had some success in upgrading the quality of canned butter concentrate for use in their field rations. The quality was not good enough to pursue commercial marketing opportunities.[37] The following year the Saskatchewan Federated Cooperatives began to manufacture margarine. The goal was to upgrade the volume content of butter in its margarine from the usual 10% to 50%, and the purpose was to improve the quality of its margarine and force other margarine manufacturers to follow suit. Failure to acquire vegetable oil seeds at competitive prices forced the Cooperatives to stop production after two years.[38]

Table 10-8
Percent Distribution of Oils Used in Margarines in Canada

	1965	1970	1975	1976	1977
Soya Bean Oil	49.4	41.7	42.5	49.7	50.8
Rapeseed Oil	n/a	29.1	36.1	31.7	33.3
Other Vegetable Oils	22.7	9.1	14.5	15.2	13.8
Total Vegetable Oils	72.1	79.9	93.1	96.6	97.9
Animal & Marine Oils	27.9	20.1	6.9	3.4	2.1

Source: Sahasrabudhe and Kurian in *Journal of Canadian Institute of Food Science Technology*, 12, 3 (July 1979): 140.

Table 10-9
Comparative Economic Efficiency

	Pounds of butter fat/acre	Pounds of vegetable oil fat/acre by soybeans
1943	46	150
1953	45	450
1954	60 (3 per worker/hour)	220 (20 per worker/hour)
1969	butter fat half that of oil fat	

Source: Canadian Edible Oil Food Industries, "Brief to Royal Commission on Canada's Economic Prospects" (1956), 26; "Soybean Growers Said Unhappy Over Compromise Edible Oils Bill," *Globe and Mail*, 27 March 1953, 5; H.L. Fowler to T.C. Douglas, 28 January 1958, T.C. Douglas Papers; and Sturjvenberg, *Margarine*, 25.

When some form of butter substitute was successfully created, errors of judgement in marketing spoiled the impact. In the fall of 1980, Canada Packers, Stacey Bros., and Dominion Dairies marketed whipped butter. Its single advantage was that, because of the added air content, the same amount would cover more bread. Marketing errors were manifold, among them failure to inform the public on how to store the product, and no price advantage because of the cost of the tub container. There were also two usage faults: difficulty in spreading it when refrigerator-cold and its odd behaviour when used in cooking.[39]

The dairy industry continues to search for ways to improve their product, particularly with regards to the problem of butter's spreadability when cold.[40] Other alternatives placed before the consumer are calorie-reduced butter (under 40% butterfat) and whey butter.[41] The latest effort to provide the consumer with an alternative is Ault Foods Limited's Lactancia Pure and Simple, which contains 52% less butterfat, 46% less cholesterol, 25% less salt and 46% fewer calories than real butter.[42] The province of Quebec allows creameries in that province to add flavourings — garlic is the favorite — and some spices such as relish, nuts, herbs, olives, pickles and dressings.[43] For the future there is the possibility of a butter without cholesterol.[44]

One major circumstance has prompted the eagerness of the dairy industry to create alternatives to butter. Margarine might be here to stay but a collapse of the whole dairy industry, a bulwark of Canadian agriculture, has been perceived as imminent if substitutes for butterfat are permitted in other dairy products such as ice cream and whipping cream.[45]

A significant opportunity to expand the use of Canadian butter when it is in surplus would be to use it in the production of margarine. Butter was used in earlier years to give margarine a texture and a flavour much closer to that of butter. Since the 1950s, most jurisdictions forbid the practice despite Norway's and Sweden's success in absorbing butter surpluses in the manufacture of margarine.

In 1979, the Ontario government set up a committee of the legislature to consider the question of permitting margarine/butter blends in Ontario. The Ontario Dairy Council was the sole supporter of the legalization of blends. Opposition came from five provincial and national dairy lobbying groups as well as from the edible oil lobby. The question was too complex for the Consumers' Association of Ontario, the Soya Bean Growers' Marketing Board or the National Dairy Council to offer a recommendation.[46] Lorne Henderson, the Minister of Agriculture, refused to move from the *status quo* to permit blends. Protection of butter was his excuse, as Nova Scotia and Saskatchewan had already moved to permit blends and butter's share of the market seemed, at least in the short run, to be on the decline.[47] Henderson ignored evidence from several countries, such as Sweden, of the long-run expansion of butter consumption.[48]

In the propaganda war waged by the protagonists, the economic components in the butter/margarine competition are frequently cited. The butter industry argued that it was the more "patriotic" and thus properly deserving of consumer support, as it supplied an income base to a very large number of Canadians who worked on farms and in the creameries. Butter apologists portrayed margarine as the national enemy insofar as it was produced by large firms using "foreign" raw materials. Margarine lobbyists could only counter with the economic arguments of low cost and versatility.

The census reports of 1951, 1961, 1971 and 1981 provide opportunities for assessing the accuracy of the butter industry's self-image. This contention used to be sound but circumstances have changed dramatically over the past thirty years. There has been a decisive switch to larger herds and greater capitalization. In 1951, 69% reported herds numbering one, two, or three to seven cows—the single largest category. Twenty years later, only 42% of the herds were in these three smallest size categories, with herds of three to seven being supplanted by 18 to 32 as the largest category. Between 1951 and 1971 the biggest category, 93 and over, had increased tenfold.

Table 10-10
Census Farms Reporting Milk Cows

No. of milk cows	1951	1961	1971	1981
1	69,742	34,535	21,493	14,002
2	64,721	30,821	12,641	—
3- 7	180,907	95,904	27,436	7,663
8-12	85,250	64,595	18,738	4,229
13-17	31,001	36,519	14,402	3,360
18-32	20,758	37,866	31,148	15,587
33-47	2,033	6,424	12,953	12,638
48-62	428	1,522	4,167	5,781
63-77	117	422	1,284	2,164
78-92	55	180	489	1,029
93 and over	56	192	567	1,426
Total farms reporting	455,068	308,980	145,318	67,899

Source: Canada, *Census*, 1951-1981.

As to total capital value a similar picture of expansion is evident. While in 1961 (the first census for which figures are available) 89.3% of dairy farms were capitalized to amounts up to $49,999, in 1981 96.8% were capitalized at somewhat over that amount. For the larger farms, a partnership of two families is becoming increasingly common.[49]

On the other hand, the image of margarine has not been built around the producer of the raw material, but around the manufacturer of the final product. There has never been the same concentration upon one raw material, and while the producer's share of the retail value of butter has

Table 10-11
Farm Data by Total Capital Value — Canada

Value	1981		1971		1961	
Under $10,000	36	(0.05%)	5,127	(3.5%)	53,258	(17.2%)
$10,000 to 24,999	402	(0.59%)	26,192	(18.0%)	134,596	(43.6%)
$25,000 to 49,999	1,778	(2.62%)	47,444	(32.6%)	88,039	(28.5%)
$50,000 to 99,999	6,850	(10.1%)	44,885	(30.9%)	27,478	(8.9%)
$100,000 to 199,999	16,078	(23.7%)	17,412	(12%)	5,609	(1.8%)
$200,000 to 499,999	27,835	(41.0%)	4,258	(2.9%)		
$500,000 and over	14,920	(22.0%)				
Total no. dairy farms	67,899		145,318		308,980	

Source: Canada, *Census*, 1961, 1971, 1981 (no figures available for earlier years).
Note: 1971 statistics have been juggled to condense 13 categories to fit the 7 of 1981; for 1961, 12 categories have been condensed to 7.

hovered between 76% and 79%, in the margarine industry the producer's share of margarine's price has fluctuated between 10% and 20%.[50] It has been the margarine manufacturer who has received the major portion of the retail price, in the range of 40%.[51]

In the World War I era, when margarine had been legalized, two firms carried on domestic production. By 1950, the year after margarine again began to be produced, there were eighteen firms in the business: one each in Newfoundland, Nova Scotia, and Alberta, nine in Ontario, two in Manitoba, and four in British Columbia.[52] The great variety of margarine legislation across Canada in these early years forced the industry to support small plants, serving small markets — at least in comparison to the American and European situations. This dispersion resulted in less efficient operations, a problem compounded by the relatively small Canadian market as well as the greater transportation requirements. The move to greater simplicity and uniformity of legislation has created a more efficient processing industry.[53] At the present time, 22 firms have reported manufacturing margarine, against 121 involved in the production of butter, seven of which produce both butter and margarine.[54]

The Canadian margarine industry operates today within a basically domestic edible-oils market. The exceptions are the imports of American soybeans for crushing, and palm, coconut, corn, cottonseed and palm kernel oil in crude form for refining and blending. These imports furnished about 15% (see Table 10.8) of total Canadian requirements in the mid-1970s and thus their prices have had some significant impact on the general Canadian price level for oils. Other influential factors are the 10% tariff on crude oils and the 17-1/2% levy on refined oils, the ban on margarine imports, and the availability and price of foreign oils in relation to Canadian supplies.[55]

The butter and margarine industries have each established a solid place for themselves in the Canadian food production spectrum, and each contributes to the maintenance of a high standard of living for most Canadians. But both are needed for this purpose. Canadian butter is no longer competitive in the world market, even if supplies were sufficient to seek foreign outlets. This conclusion was substantiated by a 1989 study commissioned by Agriculture Canada.[56]

Table 10-12
Quarterly Wholesale Butter Prices in Canada
and in Selected Countries, 1958

Butter (cents per pound)	First Quarter	Second Quarter	Third Quarter	Fourth Quarter
Canada, Wholesale Solids 1st grade, fob, Montreal	62	63	62	63
Denmark, Jobber's price for 'Lurbrand' butter	33	23	27	36
Netherlands: creamery, ex-factory Leeuwarden	49	40	34	41
United Kingdom:				
New Zealand, 1st grade, wholesale, London	30	25	28	32
Danish, wholesale, London	33	27	32	41

Source: Royal Commission on Price Spreads of Food Products, *Report*, Vol. 3 (1960), 247.

The needs of a growing, primarily urban population have outstripped the capacity of the butter industry to meet them. Margarine has had to be placed on the grocer's shelf to help meet total fats and oil needs as well as the special requirements generated by health factors.

CHAPTER ELEVEN

Margarine, Butter, Nutrition and Colour — "Holy Cow, It's Still a Holy Cow!"[1]

In nineteenth-century Canada, Canadians, even though urbanized, postulated that they were still fine citizens because of their rural stock. Those in the rural sector of society claimed they were better persons than otherwise because they lived an uplifting life — and milk, a product of rural life, was appreciated as one of the elements of their uniqueness. It was the most vital of foods, providing both energy and body-building capacity. It also came to be appreciated as a symbol of purity in spite of the massive failure of nineteenth-century Canadian farmers to produce butter at that high level of quality. Butter was *the* breadspread; butter was *the* cooking fat.

When margarine came on the scene it created an uneasiness in Canadians. The unique, the natural, was being displaced and Canadians were not comfortable with this contradiction. Margarine, by attacking ritually significant butter, came to be stigmatized as an imposter carrying with it dangers of association.[2] Margarine, beginning existence in Canada as the unnatural, had to find its place.[3]

The century during which margarine found its niche in the national economic and food spectra has also seen momentous developments in the science of nutrition. That knowledge enabled society to free itself of its prejudices against margarine and to accept it, not just as a butter substitute, but as an excellent food in its own right.

Two aspects of the nutritional history of any food must be kept in focus: the individual's concern regarding the use of a food to meet personal needs; and society's perception of what foods are necessary to meet national and even global requirements. Canadians in the mid- to late-twentieth century have come to appreciate margarine as contributing to and protecting one of the highest nutritional standards of living that any society has had the good fortune to achieve.

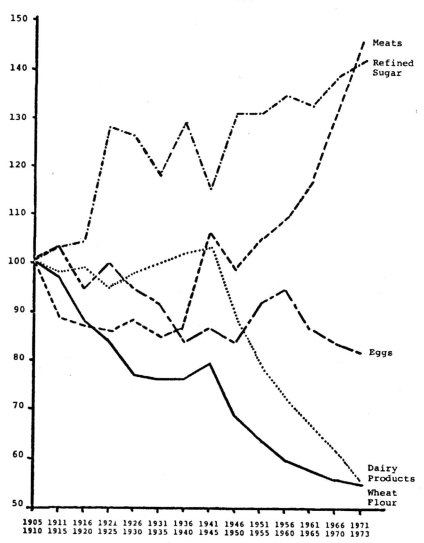

Graph 11-1
Indexed Consumption Per Capita of Selected Food Commodities, Canada, 1911-73

1905-10 = 100

Source: Food Prices Review Board, Wirick Report (1976), 8.

In the affluent four decades since the end of World War II Canadians, like most other Western industrial societies, have moved from stressing relief of the problems of under-nutrition to those related to the effects of over-nutrition. A new diversity in diet, readily available to the public, strikes at the very roots of the demand for the dietary staples of the past.[4] In 1917, the original dietary guideline took note of 131 foods, with a breakdown by proteins, fats, carbohydrates and calories. A half century later the public was presented with a list of 520 foods analyzed for nutrient value under 15 classifications: measure, weight, moisture, food energy (calories), protein, fat, fatty acids, carbohydrate, calcium, iron, vitamin A, thiamine, riboflavin, niacin and ascorbic acid.

In spite of all the efforts during World War II to achieve a sounder diet, Canadians still had a long way to go before adequate standards would exist for all citizens. In 1948, even hospital diets did not always meet the minimum needs of the patients, let alone provide the hyperalimentation which would have contributed to early convalescence and helped avoid complications and early mortality.[5] On the global scene, bleak post-war conditions were considered desperate but not likely to continue indefinitely. New strains of seeds, more fertilizer, and new techniques and machinery would enable the global community to expand food production by 20% even with no extension of cultivated land.[6]

Table 11-1
Nutritive Value, Edible Portion of 100g (1980)

	Butter	Margarine (Vitamin A added)
Food energy	722.4 cal.	725.7 cal.
	2986.6 kJ	3000.4 kJ
Protein	0.8 g	0.6 g
Fat	81.7 g	81.6 g
Carbohydrate	0.0 g	0.4 g
Calcium	23.7 mg	20.1 mg
Phosphorus	22.8 mg	16.1 mg
Iron	0.1 mg	0.0 mg
Vitamin A	93.3 RE	1076.5 RE
Thiamine	0.00 mg	—
Riboflavin	0.03 mg	—
Niacin	—	—
Ascorbic acid	0.0 mg	0.0 mg
Total Folate	2.8	0.0

Source: Robbins and Barewal, eds., *The Apparent Nutritive Value of Food* (Agriculture Canada, 1981), Tables J.1 and K.1

In 1949, the Canadian Council on Nutrition, after two years of study, issued a new dietary standard. The major new factor taken into account was the concept of body size. Suggested menu patterns referred only to butter since margarine was not yet legal in Canada. However, in the table of contributions made by each food to nutritional well-being, butter and margarine were considered as being of equal value.[7] The inclusion of margarine illustrated a paradox: a public body spending public money to investigate and report on the nutritional structure of a food illegal for use by the public.

Since 1949, there has been continuing concern about national nutritional standards and guidelines. Between 1970-1972 in particular, "Nutrition Canada," the first study of the nutritional health of an entire industrial nation, at all socio-economic levels, was undertaken. To general concern for sounder health, and an appreciation of the connection between nutrition and health, was added a concern over changing food consumption patterns by an affluent society which included "junk" foods.[8] The "Nutrition Canada" report (1974) focused on expansion of government's involvement in the upgrading of the nation's nutritional habits, in particular by implementing proper nutritional standards and educating the citizenry to appreciate their personal responsibility for their eating habits. The report did not concern itself with specific foods, let alone mention either margarine or butter.

Somewhat less official, in that they emanated from the Food Prices Review Board rather than from Health and Welfare Canada, were *What Price Nutrition?* (1975), and *A Preliminary Paper on Some Food Policy Aspects of Nutrition and Health* (Wirick Report) (1975). *What Price Nutrition?* argued that the average Canadian could eat a more nutritious diet and might do so at a monetary saving. The barrier to proper nutrition was not financial but was due to a lack of knowledge about what constituted a balanced meal.[9] The report concluded that good health and sound nutrition were inseparable. While types of food consumed, techniques of food processing, food additives and infant feeding had changed noticeably in the past hundred years, it was felt that the nutritional status of Canadians had not improved. A decline in the consumption of dairy products and wheat flour was offset by an almost identical increase in the quantity of meat and refined sugar consumed. A major increase in fat consumption resulted in a significant shortage in the consumption of calcium and vitamin D. Coincidentally, it was established that 41% of all diseases were nutrition-related, at an annual cost to the nation of $1.82 billion. It was evident that a national nutrition awareness program was desperately needed.

The publication of *Recommended Nutrient Intakes for Canadians* by Health and Welfare Canada, in 1983, represents the latest major effort by the public sector to provide Canadians with sound nutritional guidelines.[10] The thrust of this table was that Canadians needed, and could acquire, a balanced diet. It has been worked into "Canada's Food Guide," the popular format by which Health and Welfare Canada attempts to get nutritional information into each household.

Within this context of growing nutritional sophistication, butter and margarine play relatively minor, but still significant, roles. The two foods are very similar insofar as nutritional components are concerned. The contribution of butter to food energy has declined in recent decades from a high of 6% in 1963 and 1964 to 3% in 1974, while margarine's share has held almost constant at 3%.[11] Neither butter nor margarine contribute to proteins and carbohydrates. From 1960-1975, butter provided between 8-14%, while margarine has contributed between 6-8%, of the nutritive value of fats consumed by Canadians.

Of the basic fats used in the average Canadian household, there has been, between 1955 and 1980, a very steady level of total fat consumption. Within that stable picture, butter has suffered a 50% decline in per capita consumption, while margarine has enjoyed a 50% increase.

In the scientific appreciation of nutrition, milk is portrayed as our "most nearly perfect" food: high quality proteins; richness in minerals, particularly calcium and phosphorus; easy digestibility of its fat; richness in fat-soluble vitamins; high value of thiamine, riboflavin, niacin and the other members of the B-vitamin complex.[12] The only component lacking in milk to make it perfect is an adequate amount of iron.[13] Butter has shared in the accolades which have been bestowed upon milk. But, except for fat and fat-soluble vitamins, butter has none of these nutrients.

When a food like milk is understood to be near-perfect, then it is hard for the industry or scientists to contemplate how to improve on it or for the public to accept such improvements. Continuing efforts have been made to delve into the "scientific excellence" of milk and to improve on it. Shortly after World War II, Dutch scientists isolated and identified, in butter produced from milk from cows feeding on summer pasture, a previously unknown growth-producing nutrient called vaccenic acid. This butter contained seven times as much vaccenic acid as other animal fats; vegetable oils contained none.[14] The fluoridation of milk has become the project of Edgar Barrow in the United Kingdom, where "Dentamilk" is marketed as an alternative to fluoridated water.[15]

Butter has, over the years, set the standard of quality in "appearance, body, texture, spreadability, mouth-feel, mouth-getaway, flavour, and after-taste," as well as natural fat-water emulsification properties.[16] It was in the area of nutrition that margarine and butter came into conflict.

Margarine began life as an inferior food, a substitute, created to help low-income families achieve an acceptable level of nutrition. It has become an alternative to butter and has even achieved its own nutritional position as a specialty food for people with special health concerns, although here margarine manufacturers have made claims that are not completely justified.

In 1949, when Euler's work to legalize margarine was reaching its peak, margarine was considered a food equal to butter in nutritional quality: equal in percentages of protein (9.6) and of fat (81). The two foods were virtually equal in calories per hundred grams — butter, 716, margarine, 720. The natural high content of vitamin A and lesser amounts of vitamins D, E and K which butter contained were matched artificially in margarine.[17] This ranking developed as a result of scientific and technological advances achieved elsewhere.

The move by margarine to alternative, rather than substitute, food is primarily the result of the expanding knowledge of nutrition and the margarine industry's versatility in meeting these new circumstances. The chief catalyst

Table 11-2
Actual Contribution to Fat by Major Foods in Canada, 1960-75
(in grams)

Year	Wheat Flour	Sugar	Margarine	Butter	Shortening and Shortening Oil	Potatoes	Beef	Pork	Eggs	Chicken	Fluid Whole Milk	Cheese	Total Fat
1960	2	0	9	17	12	0	21	19	4	2	14	3	132
1961	2	0	10	17	11	0	21	18	4	2	14	3	131
1962	2	0	10	18	12	0	22	18	4	2	14	3	133
1963	2	0	9	19	12	0	23	18	4	2	13	3	135
1964	2	0	9	19	13	0	24	18	4	2	13	3	137
1965	2	0	9	19	12	0	25	17	4	2	13	4	136
1966	2	0	9	18	16	0	26	17	4	2	13	4	141
1967	2	0	9	17	17	0	25	19	4	2	13	4	144
1968	2	0	10	17	18	0	26	19	4	3	12	4	146
1969	2	0	10	16	19	0	26	18	4	3	12	4	146
1970	2	0	9	16	19	0	26	21	4	3	11	5	147
1971	2	0	9	16	19	0	27	24	4	3	11	5	151
1972	2	0	10	15	21	0	28	22	4	3	11	5	152
1973	2	0	10	14	22	0	24	20	4	3	11	6	147
1974	2	0	11	13	21	0	25	21	3	3	11	6	150
1975	2	0	12	12	22	0	27	18	3	3	11	6	147

Source: Robbins and Barewal, eds., *The Apparent Nutritive Value of Food* (Agriculture Canada, 1981), Tables J.1 and K.1.

Table 11-3
Percentage Contribution to Total Fat by Major Foods in Canada, 1960-75

Year	Wheat Flour	Sugar	Margarine	Butter	Shortening and Shortening Oil	Potatoes	Beef	Pork	Eggs	Chicken	Fluid Whole Milk	Cheese
1960	2	0	7	13	9	0	16	14	3	2	11	2
1961	2	0	8	13	8	0	16	14	3	2	11	2
1962	2	0	8	14	9	0	17	14	3	2	11	2
1963	1	0	7	14	9	0	17	13	3	1	10	2
1964	1	0	7	14	9	0	18	13	3	1	9	2
1965	1	0	7	14	9	0	18	13	3	1	10	3
1966	1	0	6	13	11	0	18	12	3	1	9	3
1967	1	0	6	12	12	0	17	13	3	2	9	3
1968	1	0	7	12	12	0	18	13	3	2	8	3
1969	1	0	7	11	13	0	18	12	3	2	8	3
1970	1	0	6	11	13	0	18	14	3	2	7	3
1971	1	0	6	11	13	0	18	16	3	2	7	3
1972	1	0	7	10	14	0	18	14	3	2	7	3
1973	1	0	7	10	15	0	16	14	3	2	7	4
1974	1	0	7	9	14	0	17	14	2	2	7	4
1975	1	0	8	8	15	0	18	12	2	2	7	4

Source: Robbins and Barewal, eds., *The Apparent Nutritive Value of Food* (Agriculture Canada, 1981), Tables J.1 and K.1.

Table 11-4
Per-Capita Annual Supplies of Food Moving into Consumption

Year	Margarine (kg)	Lard (kg)	Shortening and shortening oils (kg)	Salad and cooking oils (kg)	Butter (kg)	Total fat content (kg)
1955	3.67	3.95	4.40	1.09	9.34	20.09
1960	4.26	3.27	4.26	1.86	7.70	20.37
1965	3.95	3.27	4.49	2.13	8.44	19.91
1970	4.22	not given	6.94	2.63	7.17	18.78
1975	5.25	not given	7.89	3.54	5.25	19.91
1980	5.38	not given	8.75	3.65	4.51	20.39

Source: *Canada Year Book*, 1955, 1960, 1965, 1970, 1975, 1980.

for this technological revolution has been public concern about the impact of various fats on the human body — the not yet fully understood cholesterol issue.

Medical research has established a very close relationship between an elevated level of cholesterol in the circulatory system and a high incidence of heart and circulatory diseases. While the level of blood cholesterol is known to be important, a much more significant question is what regulates the level of blood cholesterol in individuals. Cholesterol is a molecule that is essential to life. In adults at least, the cells of the body are able to produce their own; the amounts present in the diet normally reduce the need for synthesis in the body.

Butter fat contains cholesterol; margarine does not, unless butter fat enters into its composition. In addition, the composition of the fat itself differs in butter and in margarine. Animal, and therefore milk, fats tend to have a high level of saturated versus polyunsaturated fatty acids whereas plant and marine oils are rich in polyunsaturates. Some of these polyunsaturated acids are known as "essential fatty acids" because they are required to sustain health in humans. For some reason that is still poorly understood, polyunsaturated fatty acids, when present in the diet in sufficient proportion, can lead to a reduction of blood cholesterol. Because in margarine the polyunsaturate/saturate ratio is higher than in butter, it is believed that consumption of margarine may be in some way healthier than consumption of butter.

The advantages of margarine in meeting health challenges lies in the fact that the food can be made from a wide range of vegetable oils, as well as animal fats (see Table 10-8). The result of this changing pattern of oil sources has led to the availability to consumers of a wide spectrum of margarines.

The advantage of using a wide range of basic ingredients is offset to a small degree by the disadvantages, in particular in the area of allergies. In meeting the dietary needs of persons suffering from an intolerance to some fatty compounds, some margarines proved better than butter, while others were far worse.[18] The butter industry has made much of the fact that its product is superior because its ingredients are relatively simple and include only milk fat, milk solids, bacterial culture, salt and (ironically) food colouring. On the other hand, margarine is a food with a much more complicated list of ingredients as listed in the federal food and drug regulations (see Appendix 8). Clinical allergists have established that some individuals may develop allergies to some of the components of margarine. This poses a problem since the formulations allowed for margarine and the source of the oils it contains may change without any alteration on the labelling of the product.[19] Another problem that has arisen in the formulation of margarine has been how to make a liquid oil into a semi-solid spread while keeping most of its healthy properties. The methods used to render margarines solid did not always yield a desirable final chemical composition. There are many classes of unsaturated fatty acids but at the moment, the claim for level of total polyunsaturates must be based on the measured level of only the most important class: cis, cis, methylene-interrupted fatty acids.

Table 11-5
Percent Distribution of Margarine Samples Containing Less than 5% to More than 20% Cis, Cis, Methylene-interrupted Polyunsaturated Fatty Acids (CCM1)

Year of survey	No. of samples	CCMI%			
		Less than 5%	5-10%	10-20%	More than 20%
1964	20	15	80	n/a	5
1976	50	20	32	14	34
1978	95	26	21	8	45

Source: Sahasrabudhe and Kurian, in *Journal of Canadian Institute of Food Technology* (July 1979), 140.

Another concern was the presence in some vegetable oils of undesirable fatty acids which had been shown to have deleterious effects on health; e.g., erucic acid in old cultivars of rapeseed. This problem has been largely circumvented by the development of new cultivars that do not show these undesirable properties.

The latest medical research into the impact of fatty acids has led to strong pressures upon the federal government to update its regulations regarding the ingredients in margarine. The thrust has been to require no less than 5% of linoleic acid, an essential fatty acid which is a precursor of molecules that are important in blood clotting, protection against bacterial infections and other traumata, and many other functions; no more than 25% saturated fatty acids; and no more than 40% of a combination of trans fatty acids and saturated fatty acids.[20] However, these standards have been attacked by some, working at the behest of the Dairy Bureau of Canada, as simply falling short of achieving levels effective in upgrading Canadian diets.[21]

According to the "Nutrition Canada" survey, the average Canadian diet contains about 45% of its energy as fat. The prudent diet promulgated by

Health and Welfare Canada should contain less of the energy as fat, but at least 3% of the energy should come from polyunsaturated fats of which vegetable oils and vegetable oil-based margarines are the richest source. Because food is plentiful in Canada, and because a diet based on the Canada Food Guide is believed to provide all the necessary nutrients, the nutritional advice most suited for Canadians is to eat moderately, and to choose from a variety of foods.

Greater knowledge of the human body's nutritional needs has generated a wide range of margarines, giving that food a real advantage over butter in the marketplace. Consequently, the cost factor has been replaced as the chief element in consumer preference. With the passing of time, the colour factor has similarly lost its significance in most provinces.

The colour factor has played a many-faceted role in the history of the butter/margarine competition. The government's regulation of colour was grounded in the attitude of Canadians towards that factor. It seems clear that the issue of colour, a direct product of the butter/margarine competition, gradually grew into a myth which influenced public-sector policy. It also became an important protective weapon in the strategy of the butter forces. Out of perfection (milk white) to regal gold, butter sets the colour standard.

The initial 1886 debate over the role of margarine in the Canadian marketplace noted the colour factor in a manner suggesting that colour was one of the standards by which the quality of butter was judged. The better margarines could match the standards of the better quality butters in texture, appearance, flavour and colour. By World War I, it was taken as axiomatic that Canadians expected their bread spread to have a yellow colour and few consumers would want to eat margarine unless it was coloured to match butter.

In the decades since, provincial government regulations have, to varying degrees, helped protect the position of butter in the marketplace. Today Ontario and Quebec remain the only Canadian provinces with any regulatory controls other than those related to public health, and the chief operative control is the colour factor. It is acknowledged that protection of butter's position in the marketplace is the sole reason for this regulation.

The fight has centred on the narrow part of the Lovibond tintometer scale which has, in modern time, become the range set aside for butter. In technical terms the range is between 1.6 degrees and 10.5 degrees of yellow or of yellow and red combined. Butter offered for sale in Canada may range from 3.5 to 6.7 on the same scale, the average being between 4.0 and 4.5. June butter, which ranges between 5.0 and 6.5, could generate some sales resistance.[22]

Since the legalization of margarine in 1949, the colour element has continued to play an important role. In an era in which there has been a great expansion of the number of manufactured goods, it is economically important to establish, in the consumer's mind, the distinctive nature of a particular food. Thus the butter industry has, as part of its merchandising efforts, come to stress a standardized yellow, because of its distinguishing character and also because it is perceived by consumers as part of the appetizing and decorative characteristics of butter.[23] In off-colour seasons, when dry winter feed does not lead to a distinctly yellow product, natural vegetable colours such as beta carotene and annato are used.

This era also saw a broadening of suggestions as to what colour should be adopted for margarine. Among these were chocolate, something like peanut butter,[24] red and white stripes, blue when made from fish oil or brown when the basic ingredient is soybean;[25] also deep green,[26] purple,[27] or orange.[28] Canadians were only imitating actions outside Canada. New Jersey once required pink margarine, while Germany proposed to impose the brown colour of the oak wainscotting found in the Reichstag.[29] The degree of silliness which this matter reached is shown in the 1952 debate in British Columbia, during which a butter supporter facetiously maintained that to force persons to mix the colour into margarine at home was of social benefit because the exercise made them healthier![30]

Each side marshalled a wide range of arguments to help promote and protect its product. For butter's side of the case, the suggestion that butter producers have a "moral" right to the yellow colour[31] substantiates the continuing existence of the myth surrounding the colour yellow. The protective impulse is well illustrated by the *Montreal Star*: "The *Star* has always believed that butter producers should be given all the protection which the distinctive identification of margarine as margarine can be made to provide."[32]

Arguments concerning the desirability of a distinctive yellow colour for butter have been used on a variety of occasions in different political jurisdictions in Canada. The continuing confrontation over this issue in Ontario offers a synopsis of such contentions. The arguments centre on consumer protection from fraud; easy identification of margarine by persons subject to allergic reactions to chemicals in margarine; and prevention of "further cannibalization" of Canada's dairy industry.[33]

With regard to fraud, "the margarine industry alchemists' [!] ideal is a margarine that is a perfect replica of butter."[34] Success has been so great that only colour distinction remains to prevent the consumer from being readily deceived. This deception was much more possible to carry off in public eating places than in home use.

In the second place, colour identification was important for persons susceptible to allergic reactions to certain chemicals which could be used in margarine, or even for those suffering from allergic reactions to butter. Colour distinctiveness would permit easy product identification even when the wrapping had been removed.

Third, the "cannibalization" factor was primarily an economic matter. Any increase in margarine consumption would result in a countervailing decline in the use of butter. The butter industry, which used the single largest proportion of all the milk produced in Canada, depended upon the prosperity of dairying, of Canadian agriculture, and (through the multiplier effect) the prosperity of the country as a whole. Studies undertaken by the Dairy Bureau of Canada in 1977 and 1980 concentrated on the attitudes towards butter and margarine in three different parts of Canada — in the

French-speaking parts of Quebec (462 samples), a recent convert; British Columbia (254 samples), in which colour restrictions had not existed for some years; and in Ontario (889 samples), where colour distinctiveness was still the law. On the basis of both eye appeal and taste, Ontarians made the greatest distinction between butter and margarine. They also expressed a much stronger colour preference for butter. In Ontario both incidence and frequency of use of butter were greater than in the other areas surveyed. The conclusion of the Dairy Bureau was that colour distinctiveness provided a significant advantage to the butter industry and was required to offset the price advantage held by margarine.[35]

To counter the butter interests' contentions, the margarine lobby has had to exert much effort. The Institute of Edible Oil Foods (IEOF), founded originally to fight both the colour ban and the 12% federal sales tax,[36] has had since 1971 as its primary *raison d'etre* the removal of colour restrictions.[37] The margarine industry's argument is that there are no economic or social reasons for maintaining any restrictions other than those related to pure foods regulations. The IEOF has set out six basic reasons for its position:

(a) consumer preference: colour restrictions discriminate against persons opting for margarine for economic, tax, or dietary reasons;

(b) other regulations provide sufficient protection from fraud. Approximately 60% of margarine sold in Canada in 1981 is in the soft, spreadable form in round, plastic tubs. This form obviously imitates butter in no way.

(c) margarine manufacturers are being discriminated against by being required to produce two different coloured spreads and, thus, a double inventory.

(d) removal of colour restrictions in Ontario would have no marked impact on any possible shift of consumers from butter to margarine. Thus the dairy industry, which is recognized as a very important sector of the economy, would not be hurt. The dairy industry is not now able to supply all Canadian dairy needs. Even if the worst scenario of a 4% shift were to occur, the national milk supply quota would be affected by less than 1%; the dairy industry has argued that the shift could easily be in the magnitude of 8 to 13%.[38] (A second analysis supported the butter industry's conclusion: the colour differentiation is annually worth $22 million or $3,000 per dairy farmer in Ontario.[39])

(e) maintenance of the colour ban is inconsistent with the Ontario government's policy of supporting the edible oil seeds sector of Ontario agriculture.

(f) the IEOF position has the support of a wide sector of the Ontario economic spectrum: consumers, farmers, and manufacturers.[40]

Political considerations remained. In October, 1976, the IEOF presented to William Newman, Ontario Minister of Agriculture and Food, another in its series of submissions in support of ending provincial colour restrictions. Using a statistical analysis of Ontario constituencies, the parties holding each constituency, and the presence or absence in each riding of margarine manufacturers, soybean growers, and strong contingents of the Consumers' Association of Canada, the study demonstrated that:

(a) in two of the four major dairy areas of Ontario the Progressive Conservative party held a substantial majority, 69% and 80% of seats over the Liberals' 8% and 13%, and the New Democrats' 23% and 7%. In the remainder of Ontario the PCs held only 38% of the seats;

(b) in the rural ridings in which soybeans were a major crop the PCs held only 42%, Liberals 58%, and NDP 0%;

(c) in the ridings in which the Consumers' Association of Canada presented a vital factor the PCs again held a minority 35%, Liberals 26%, and NDP 39%.

(d) in districts in which the plants of the edible oil manufacturers were located the PCs held 22%, Liberals 22%, and NDP 56%.

The conclusion drawn by the IEOF was that the pro-butter policy regarding colour was working to the party's advantage in only a small part of the province and thus contributing in a "large measure" to the minority position of the PC government.[41]

One issue does not usually determine the fate of governments, but the continuation of the ban in Ontario demonstrates the political clout of the dairy industry, and that there are millions of dollars at stake when consumers express their preference in the marketplace, a preference influenced to some degree by the colour factor. This economic and political strength generated a sympathetic response from the provincial PC governments. In 1960, the Leslie Frost government refused to bend under the pressure of the Ontario Margarine Committee, a group organized to fight the colour ban, which presented the government with a petition signed by 30,000 persons.[42] The succeeding William Davis government also remained firm. After a major confrontation in caucus, the Liberal David Peterson government fell into line.

No other food has enjoyed such protection in the annals of Canadian food history. The myth factor helps explain why Canadians were willing to extend to butter such unique support. It has been one of the factors used to carry on the conflict that rages between producers, as each attempts to protect or expand its market share.

CHAPTER TWELVE

Marketplace Competition[1]

From almost the first days of its existence in Canada, the butter industry has seen itself as under siege by an external enemy. Historical events since 1886 have served to intensify this feeling. In the case of margarine, the bans of 1886 and 1923 and the discriminatory legislation since 1949 have generated a wariness towards the dairy industry and the federal and provincial governments, a sense of suspicion intensified by margarine's image as a second-class product. Butter has always provided the gold standard which the margarine industry has felt obliged to meet.

In the new competitive environment the butter industry responded by worrying about the future and seeking new legislative protection: ban foreign edible oils used in the manufacture of margarine, institute a quota system in order to control the quantity of margarine sold, and increase the federal sales tax.[2] If price was the prime factor in stimulating margarine sales, then act to wipe out the price differential.

The margarine industry met the challenge in two ways: product improvement and product promotion. In North America, margarine had been changed little since the addition of vitamins in the 1920s and the shift from foreign to domestic oils in the 1930s.[3] Depression and war were not conducive to the development of new products. It would require the abolition of American federal restrictive taxation in 1950 to stimulate major improvements in the United States and, as a spin-off, in Canada.[4] Heightened social aspirations and incomes of postwar Canadian society left most consumers willing and able to purchase better food. To gain and maintain their market share, margarine manufacturers were compelled to provide a product of better quality, but often of higher price. Such an approach made inroads into the butter market possible.[5]

The first margarine to reach Canadian store shelves was produced by a British Columbia firm, Associated Producers, in Burnaby in 1948.[6] The national firms were led into the marketplace by Canada Packers. Other national firms, such as Burns and Swift Canadian, hesitated because of the high cost of equipment and the scarcity of oils, as well as the lingering uncertainty of market success until the judicial appeals had been seen through the courts to final decision. The evident hostility of some provincial governments towards margarine also led these firms to demand assurances from provincial attorneys-general before acting.[7]

The first margarines to reach the national marketplace were Margene produced by Canada Packers, and Nucoa from Best Foods. Their appeals to the public stressing quality of flavour and economy, the manufacturers offered a product which was spreadable at room temperature and had been fortified with 16,000 units of vitamin A per pound.[8] While some argued that it was indistinguishable from butter, the more discerning palates found little resemblance in either texture, colour or taste — "like a mixture of mayonnaise and cream cheese."[9] By September 1949, Margene had been improved to make it spreadable, either warm or cold.[10] A year and a half later, foil wrap, a major improvement in packaging, provided much greater protection of flavour. These innovations were apparently successful, as Canada Packers could claim that Margene was the nation's largest selling margarine.[11]

The entry of the international firms intensified the pattern of product improvement. Jelke's Good Luck brand, in 1951, offered twin bar packaging and the use of either wafers or the new plastic bag for working in the yellow colouring.[12] In the fall of 1952, Kraft offered a margarine "spreadable even when ice cold" as a result of blending two hydrogenated vegetable oils.[13] Kraft's Parkay also achieved greater nutritional quality because it contained a lower level of saturated fatty acids — a major selling point to the present day.[14] Standard Brands' Fleischmann's margarine marked the last of the major product improvements in this initial period of production. Made from corn oil and thus having a high level of polyunsaturated fat, it offered a response to the growing national concern over cholesterol. Standard Brands boldly extolled its product as "the greatest margarine discovery in 50 years."[15]

The industry now had a product which it felt was more healthful than butter, at least in the context of prevailing standards. As a result, it had achieved security in the marketplace, had overcome its inferiority complex, and had gained a position from which it could attack butter's market share even more strongly. Innovative efforts did not stop. Soft margarine achieved public acceptance, something that other specialty margarines, such as liquid, unsalted, and low calorie, had not. The expanding research into new oils, such as sunflower (with even fewer saturated fats), has since led to such margarines as Becel.

In the post-World War II era, economy was the prime reason for purchasing this new food. In anticipation of the arrival of margarine, the *Brantford Expositor* had informed its readers they would be able to purchase one pound of margarine and two quarts of milk for the price of one pound of butter.[16] In 1948, butter sold at an unprecedented high price of 69.8¢ per pound wholesale. The arrival of margarine in the marketplace generated strong downward pressure on price and upward pressure on supply as butter producers moved to make sure that they were not caught with a glut.[17] The wholesale price of butter declined by 7.9¢ per pound in 1949 and a further 4.0¢ per pound during 1950 before climbing again in 1951.[18] On the retail market the first margarine was sold in Vancouver for 53¢ per pound.[19] In September, 1949, when the initial flurry over margarine took place, a store in Kitchener-Waterloo advertised the two spreads on sale: butter at 59¢ per pound and margarine at 35¢ per pound.[20]

Product promotion became the second major marketing tool. Uncertain how long the ban would remain in limbo after 1917, the industry had had little incentive to advertise its wares, improve them, or work to create a permanent body of consumers. After World War II, growing signs that margarine would remain legal, at least in some parts of the country, stimulated the margarine manufacturers to take steps to create a permanent and expanding market. Newspapers and magazines disseminated the new message. Other, less customary steps were also taken. For example, on the day before Christmas, 1948, an "introduction party" was held in Toronto to which the press, dietitians, and home economists were invited.[21] Such actions generated a flood of support from the media. *Maclean's* not only welcomed margarine editorially, calling butter as costly as "gold leaf," but also published an article which provided its readers with a history of the food, set forth its ingredients and how it was produced, and told its readers how to pronounce the name—"use a hard 'g,' as in Margaret."[22] The *Canadian Home Journal* informed its readers that, of the two options (hard "g" or soft "g") the preference was for the first pronunciation: the periodical also ran an article giving its readers a few recipes in which either butter or margarine could be used.[23]

The aggressiveness of the margarine industry, first seen in the development of new product lines, was also manifest in the second area of attack. An appreciation of market segmentation led to the development of three unique marketing strategies, in order to exploit all segments of the Canadian consumer range efficiently.

The most secure segment of the market which margarine acquired is the one that purchases the food because of its low price, and is little concerned with quality—the large, low-income families who can afford nothing but the cheapest foods. These cheaper brands are sold under the vendor's name, e.g. "Schneiders," or under a brand name such as "Mom's," by Monarch. Since price is the sole criterion of choice and the purchasers are not concerned with any other aspect of the product, these margarines are not advertised.[24]

The second group of purchasers is concerned not only with value, but also wants a good quality product at a reasonable price. The consumer profile of this group is similar to Canadian society as a whole. Margarine is purchased for a variety of reasons: price, flavour, convenient packaging, and for health reasons, such as diet restrictions. Although exact figures are not available, it is estimated that this market is the largest. The products offered are rarely advertised; instead, coupons, store displays and price adjustments are used to promote them.[25]

Paradoxically, the brands of margarine produced for the smallest portion of the market are the object of the most aggressive promotion. The premium brand market is characterized by above-average incomes, small families who consume little margarine, and persons of all age groups who purchase foods with a great deal of concern for nutrition and health. Margarines such as Fleischmann's and Becel are promoted on grounds of health and taste, not economy, since the price is often comparable to that of butter. Table 12-1 lists the average retail price per pound of butter and of the three categories of margarine, and illustrates the point that the premium brands are purchased for reasons of quality of product and not price.[26]

Table 12-1
Average Retail Price Per Pound

	Butter	Margarine		
		Premium	Value	Economy
1972	.74	.60	.39	.30
1973	.76	.62	.41	.32
1974	.87	.83	.61	.52
1975	1.07	.97	.78	.63

Source: Institute of Edible Oil Foods, *Submission to the Ontario Government* (1976), 5.

If the premium market is so small, why are these products the subject of virtually all the advertising efforts? The principal reason is that despite the change in Canadian culture over the past hundred years, margarine has retained the image of a second-class product. Countering this image with product innovation and promotion of the premium lines also helps the product development of the economical brands. Innovation and advertising have also aided the industry to overcome its inferiority complex.

Margarine promotion work in Canada has gone through three distinct phases: introductory, followed by an offensive on the health front, and subsequently a counteroffensive on that same front. The Canadian market provided some special factors which made more difficult the task of disseminating the margarine message. Producers were not solely Canadian. Canadian firms, and British firms such as Unilever, had to follow the lead of American firms, such as Kraft, who were reacting to their American experience. All rivals then began to spend heavily on advertising and other promotions in a Canadian market in which the consumer was less ready than the American counterpart to buy a new product solely on the ground of novelty. The impact of this factor was compounded by the confusion generated by the variety of provincial restrictions, by the broad expanse of Canada, and by the bilingual/bicultural fact which created two distinct markets.[27]

To marshal the margarine forces, growers of oil seeds, manufacturers of margarine and producers of vegetable oils grouped together in the Institute of Edible Oil Foods. One of its initial actions was to mount an annual conference on the economic and nutritional significance of margarine. In 1954, for the first of these meetings, Senator Euler was brought out of retirement from the margarine controversy to review his long fight and to urge renewed efforts to defeat the Ontario colour ban.[28] For the second, such luminaries as C.A. Massey, president of Lever Brothers, urged public pressure, especially on the provincial governments, to have their discriminatory legislation eased or abolished.[29]

This initial phase, lasting a half dozen years, also concentrated upon introducing the new food to Canadian consumers through an extensive newspaper and magazine campaign. Several themes were highlighted: introduction of a product already possessing great consumer acceptance in the United States (Nucoa, 1949), economy (Margene, 1949), ease of spreading (Margene, 1949), new colouring method (Good Luck, 1949), and flavour (Allsweet, 1950; Good Luck, 1951; and Parkay, 1952). A strong current of defensiveness ran through the advertisements: margarine had been fortified with vitamins and therefore was not nutritionally inferior to butter;[30] kneading the bag to work in the yellow colouring was "fun,"[31] and Allsweet, at least, was desirable because "it's *guest quality*" margarine. The industry also stepped on its enemy's toes by arguing that the milk in Allsweet made the difference in flavour quality[32] and by suggesting that Good Luck was "sweet-churned."[33] A third hint came when coloured advertisements were used. Since only one shade of yellow was available to the publisher of advertisements in colour, margarine had the same appearance as its competitor. Recipes were also frequently included to demonstrate that margarine was as good as butter and could be substituted in cooking and baking.[34]

Giveaways and contests formed another category of gimmicks used in advertising the product. Unilever concentrated on clothing by offering $1.20 off the price of two pairs of nylons for including the end flap of the Good Luck package. The stockings came in six sizes and two shades and "would serve as excellent Christmas gifts."[35] The company also ran a contest in which eight first prizes were women's Easter outfits up to a value of $200 each.[36] Kraft stressed cars as first prizes in each of three contests in early 1953. A $200 cash bonus would be paid, in addition to the Ford Victoria, if the entry was accompanied by two Parkay package end flaps.[37] Swift's promoted Allsweet by offering a first prize of $1,000 and other prizes to a total of $4,000 as well as redeeming each end flap for two cents. The offer was made to church congregations of any denomination in the country.[38] Through a separate program, Swift also offered discounts on silver flatware.[39]

The margarine industry spent the huge sum of $800,000 on advertising in 1949.[40] Canada Packers attempted to ascertain whether they were getting their money's worth by enclosing a questionnaire in each package of Margene placed on grocer's shelves in Ontario, New Brunswick, and Nova Scotia. The almost 22,000 replies to the four questions let Canada Packers hear what it hoped to hear: that users of Margene were nearly unanimous in their satisfaction with the food as a bread spread, and only slightly less so with it used on vegetables and in baking.[41]

By the middle of the decade, the volume of advertising declined, indicating that the goals had been achieved. Consumers were now acquainted with the new product. In February 1950, the monthly quantity of margarine produced in Canada was, for the first time in history, greater than that of butter: 8.6 million pounds to 8.5 million pounds. February was traditionally the month of lower butter production in Canada's cycle, as determined by the climate. Margarine still had a long way to go to catch up to butter on a yearly basis, or even in its peak production month of June, 1949, when 41.1 million pounds had been produced.[42] Margarine and butter both increased total production[43] although butter consumption did decline on a per-capita basis. In the booming economy of the 1950s and 1960s, both the market and the economy had expanded rapidly. Margarine manufacturers were able to increase sales without expensive promotional campaigns. Relatively small efforts to introduce new products, such as Fleischmann's in 1960 and Blue Bonnet in 1967, were exceptions.[44]

The 1960s were the lull before the storm. The knowledge of the link between heart disease and saturated fatty acids had been the catalyst for the industry to create a better product in the 1960s. In the next decade, this knowledge was used to sell new margarines.

Attempting to counter the Canadian margarine industry's efforts, the butter forces faced several real problems: lack of unity of organization and purpose, and a shortage of funds. Although the national organizations for agriculture, and more specifically dairying, had existed since before World War II, they simply could not bring their constituents together as solidly as the Institute of Edible Oil Foods could, especially in the area of monetary commitment.

To counter the arrival of margarine the butter interests first appealed the Supreme Court's ruling to the Judicial Committee of the Privy Council, an especially attractive move since the federal government was willing to provide some financial assistance. However, this attempt to maintain the status quo quickly failed and the butter industry had to meet margarine head to head in the marketplace.

The efforts centred on refuting claims made by the competition, organizing a boycott of margarine, gaining further government support, promoting the consumption of butter, and creating new products. The initial attraction of margarine for the majority of Canadians had been the cost factor. In rebuttal, the butter people argued that the savings would be minuscule, $1.30 per family per year: how this figure was arrived at was not made evident.[45] Such savings, they argued, were easily offset by the costs to the nation of the negative impact on Canadian agriculture. J.H. Duplan, president of the National Dairy Council, calculated that the Canadian economy had lost over seventy million dollars in 1949 alone, and that losses would be greater in succeeding years should margarine remain in the marketplace. He broke down his calculations as follows: (a) reduction of $50 million in purchasing power for Canadian farmers; (b) unnecessary expenditure of $10 million of badly needed American currency for imported vegetable oils; (c) investment of $10 million by the federal government to buy up a surplus of twenty million pounds of butter; (d) the [unstated] cost of replacing the milk production of 250,000 cows sold by Canadian dairymen.[46]

H.H. Hannam, of the Canadian Federation of Agriculture, argued that Canadian farmers, in seeking reasonable price levels, were only aiming to acquire the "kind of security" that all other Canadians hoped for.[47]

The butter interests also pounced on the margarine people's move to have colour restrictions removed. If yellow margarine were to be permitted, then very little white margarine would be available. In the U.S., yellow margarine sold for 12 to 25¢ per pound more than the white, even though the colouring cost only 1/10th of a cent per pound.[48] The margarine producers were thus accused of campaigning for new circumstances which would undermine the validity of their initial argument for the legalization of margarine — lower cost. These arguments did not win the national media attention that the economic contentions of the margarine lobby did.

The idea of a margarine boycott was raised immediately after the Judicial Committee of the Privy Council upheld the Supreme Court's ruling in 1950. Farmers were asked to stop all purchases of margarine, while in a less direct manner it was suggested that all Canadians ought to support policies which were in the national interest.[49] Such a step stood no chance of success, even with rural Canadians. When farmers could sell one pound of butterfat and buy two pounds of margarine, many did so. It was common practice to use butter at the table and margarine to replace lard or shortening for cooking purposes.[50] Dairy farmers proved no more loyal to butter than any other farmers.[51] Glaring inconsistencies clearly emerge from a review of their reaction to margarine. One lady from an Ontario dairy farm was prone to strong anti-margarine rhetoric, while secretly using it for cooking purposes. A Prince Edward Island father refused to permit margarine in his home, yet he did not object when his daughter served him margarine in her home.[52]

In 1958, a survey of Ontario consumer preference was carried out. While virtually all Canadians used butter at the table, a sizeable proportion must have used each food on occasion. Comparing the results of another survey taken seven years later, of the use of margarine alone, reveals a trend of increased complexity of the pattern of use.[53] Reasons given for the choices were as follows: butter was preferred because it was better for table use (whatever that might mean), it had a tastier flavour, and was nutritionally superior. Margarine preference was based on lower cost; preferability for cooking purposes; and health considerations.[54]

Continuing its defensive line of thought, the butter industry strongly urged further government protective action in the form of a tax to bring margarine's price in line with that of butter and the establishment of import controls or taxes on foreign raw materials.[55] Huge international combines "were the enemy and a thriving Canadian dairy industry was of greater value to a prosperous Canada than rich international companies."[56]

Not all perceptions were negative. George Unwin, a farmer, and president of the Saskatchewan Federated Cooperatives, argued the case of greater versatility. Margarine was to be allowed for those who needed it for economic reasons as long as fraud was controlled. The dairy industry would be better

Table 12-2
Butter and Margarine Use in Ontario, 1957 and 1965

Major Uses	Percent of Butter Users 1957	Percent of Margarine Users 1957	Percent of Margarine Users 1965
Table	94.5	28.8	47
Cooking	39.6	51.5	59
Frying	14.9	24.2	n/a
Baking	12.6	34.09	61
Other Uses	2.7	6.06	n/a

Source: Ilori, Master's Thesis, 1958, and *Canadian Consumer Survey*, 1965.

off working in a larger fluid milk market. Saskatchewan agriculture would also benefit from a chance to raise oil seeds locally.[57]

Product innovation was also attempted, but on a minor scale when compared to margarine. The reason given for dropping the project of marketing a dairy spread at the same price as margarine illustrates the paradox inherent in the situation as well as a narrowness of mind: such a dairy spread was perceived as offering as much competition to butter as to margarine.[58]

Lastly, an effort was made at product promotion. In September 1949, the directors of the Dairy Farmers of Canada resolved to begin a national sales promotion. This activity formed half of the effort to which the Dairy Farmers committed themselves at their annual meeting in February, 1950. Every effort would have to be made to offer the best products possible at the lowest price. The protective impulse showed itself again in that the advertising campaign tried to get across to Canadians that low-priced dairy products were not conducive to a sound national economy. "Purchasing power in the hands of the producer was just as important as full employment and fair wages for labour."[59]

In 1951, a dairy coordinating advisory committee was set up to plan a course of action. Unfortunately, little is known about how this group operated, or even about its membership. Apparently it levied a charge of one cent per pound of butterfat or milk equivalent on cream or milk marketed during June 1950. (June is usually the month of peak milk production.) These funds were used to finance a series of advertisements in newspapers, and magazines and on radio. This attempt was the first at such a promotion and its theme was "It's Better with Butter."[60] The industry exploited its myths, concentrating on the unique "goodness of flavour" imparted by butter when used in cooking, and offering ideas on how to use the whole range of milk products to help achieve better "winter health"[61] (whatever that may be). In this manner the butter industry began its latest battle with margarine on a considerably lower key than the massive attack mounted by its adversary.

In the 1960s, the Canadian margarine industry mounted a health offensive. It was part of a wider North American effort, which one American consumer advocate, *Consumer Reports*, termed as "one of the most unprincipled food promotions in the past quarter century.... They have promoted a staple food as though it were a drug."[62] The industry, on the defensive since 1949, had apparently decided to move to the offensive. During the fifties and sixties there was growing public concern over cardiovascular diseases. Evidence at hand, highlighted by such studies as Dr. Ansel Keys (1970) "Coronary Heart Disease in Seven Countries," established a correlation between the consumption of animal fats and the incidence of mortality from coronary diseases.[63] The margarine industry, perceiving itself at war in the marketplace and still suffering some effects of its old inferiority complex, exploited the fears these studies created in order to sell its product. Margarine was promoted as if it were an all-purpose remedy. Advertising adopted health slogans, and the implicit message was that butter was poison. Tangential efforts were made to remove the federal sales tax and provincial restrictions, primarily those concerning colour.

To help carry on the new offensive, the Institute of Edible Oil Foods employed Scott-Atkinson, Only, International Ltd., public relations consultants. Using the services of George Caldwell, the new IEOF president, and Jane Cummings, director of consumer services for the Institute, who had done excellent work during the introductory period, the margarine forces advanced to the attack, with a 40-second TV newsclip; a 100th anniversary party for margarine; a 13-minute film entitled "Strangely Colored and Punitively Taxed," to parallel a booklet by the same name; interviews of Caldwell and Cummings on radio and TV talk shows, and the customary news releases to the media leading to newspaper articles.[64] In the estimation of David Scott-Atkinson, the IEOF was too cost-conscious in its efforts, allocating $3,000 per month rather than the $10,000 he believed would have accomplished the task properly.[65] Even so, the offensive brought about the end of the federal sales tax and some curtailment of colour restriction in Alberta and Saskatchewan.

The prime thrust of the offensive was to win consumers to margarine by exploiting the health issue. While all the major manufacturers used the polyunsaturate/health theme, Standard Brands Limited capitalized on it most thoroughly. Over the next decade, Canadian magazines and television screens presented a tall, dignified, white-haired gentleman, at his doctor's order, riding a bicycle, climbing mountains, doing push-ups, and, of course, eating sensibly by including margarine — Fleischmann's, in particular — in his diet. If the viewer did not do likewise, it appeared that his days were numbered. Using slogans such as "Fleischmann's Takes Care"[66] and "You Are What You Eat"[67] the company stressed the benefits of polyunsaturates and urged Canadians to eat "sensibly." Standard Brands showed its aggressiveness by mailing doctors a coupon entitling the physician to one free pound of Fleischmann's product. Also included in the mailing was a large number of other coupons offering a 30-cent rebate; the coupon included a space for the doctor's initials and a space for the name of the patient to whom the margarine was being recommended.[68] The Canadian Medical Association noted that the action violated its code of ethics, which forbids the advocacy of any product.

The health theme was an effective weapon: sales of margarine rose while those of butter declined. Total consumption of butter and margarine had steadily declined from the 1963 peak of 28 pounds per person. Margarine consumption increased from 9 pounds per person in 1970 to 12 pounds in 1976; butter use over the same years declined from 15 to 11 pounds. These patterns have held into the 1980s. In 1974, for the first year in Canadian history, margarine production outstripped that of butter. Margarine has continued to be produced in greater quantity every year since then. After almost one hundred years, margarine was winning the war.

This shifting balance led the Dairy Bureau of Canada to mount a massive counter-offensive in July 1978. Lending urgency to the Bureau's efforts, a major publicity and education program was begun in January 1978 by the federal Department of Health and Welfare, aimed at urging Canadians to eat less fat, in particular saturated fat. The government campaign was prompted by the Fraser Mustard report of January 1977, which strengthened the argument that coronary diseases and the consumption of saturated fats were closely related, and any attack on saturated fats automatically included butter.[69]

The anti-margarine campaign used the full range of media facilities. Advertisements fell into two general categories. One series, implicitly anti-margarine and based on butter's close relationship to the agrarian myth, concentrated on the fact that "there is no substitute for the good taste of butter,"[70] and that it was a healthy food.[71] Using slogans such as "Butter Makes It Taste Much Better Naturally"[72] the Dairy Bureau set out to reinforce butter's position in the Canadian mindset.

A major part of the campaign presented executives from such large food companies as Sara Lee, Green Giant, and McDonalds, extolling the virtues of butter. Brian Homer, president of Sara Lee, stated in one advertisement: "I think the brand name of our Original Butter Recipe Cake line tells you how important Sara Lee believes butter is when it comes to achieving good taste."[73] On at least one occasion, such an advertisement was placed in *Chatelaine* alongside an article on dessert recipes in which butter was the only fat noted as an ingredient.[74] The Sara Lee advertisement was intended to drive home the view that people in the food industry accepted that butter was not only good for you but also good for business.

The explicitly anti-margarine side of this campaign by the Dairy Bureau of Canada was harsh, direct, and harked back to basic values. The slogan was "Butter can be made purely and simply. Imitations can't" — implying that margarine was an artificial, second-class product. One advertisement contrasted ingredients and manufacturing processes. Butter was made from milk which was separated and churned to produce a wholesome "naturally golden yellow" spread (never mind the colour additives during periods of pale butter production). Margarine, on the other hand, was manufactured from vegetable oils and in some cases from fish oils, lard and tallow. These oils have to be bleached; deodorized, hydrogenated or intensified; and emulsified; vitamins, flavouring, colouring, and, finally, preservatives are added, and the preservatives may include an antibacterial agent or an antioxidant. To drive home the point that margarine was artificial and inferior, butter was pictured in its traditional one-pound brick form, and margarine in a sterile plastic tub.[75]

Less orthodox parts of the butter campaign were heavy advertising in medical journals, pointing out that butter and margarine contained an equal 72 calories per 100 grams, and suggesting to doctors that margarine was not a better choice for weight-watchers. To counter the margarine advertisements in these same journals, doctors were urged to re-examine the scientific literature on the relationships between butter, cholesterol and coronary disease.[76] The medical profession was also approached by the margarine forces in ways that further demonstrated the intensity of the fight. On other fronts, Harvestore Systems Dealers of Ontario, manufacturers of sileage storage buildings, joined their economic compatriots by printing cards that would be left in restaurants to thank the management for providing butter, or to express disappointment at having been served "look-alike" imitations. A new wrinkle in advertising came with the promotion of "the first *real* alternative to butter" — whipped butter.[77] In 1985, the Dairy Bureau of Canada laid a complaint with the Department of Consumer and Corporate Affairs. It accused Monarch Fine Foods Ltd., the largest margarine producer in Canada, of violating the federal Food and Drug Act which forbids the promotion of food on the basis of health. Monarch had included in its Becel advertisements "the fight against heart disease begins with proper diet. We think you should know that Becel Margarine contains 55% polyunsaturates and 25% saturates. . . ." The department reacted by going the next step in its schedule of warnings to include a mention that court action might be taken if any further violations were to occur.[78] Nabisco was accused of a similar violation of the Act when it offered to redeem every coupon of Fleischmann's for a 10¢ donation to the Ontario Heart Foundation. The advertisement was perceived as encouraging consumers to "join the fight against heart and stroke."[79] This tactic only duplicated that of the IEOF, five years earlier, when it lodged a complaint with the Advertising Standards Council of Canada against the butter industry's attempt to play upon "widespread misunderstanding" by the public of the role and nature of additives.[80]

Government ostensibly supported both sides of the struggle. While the Canadian Dairy Commission, under the Department of Agriculture, spent $4 million to support butter consumption advertisements, the Department of Health and Welfare was admonishing Canadians to eat less animal fats. Dr. Keith Murray, head of the federal Nutritional Sciences Bureau, noted that this was not a case of the left hand not knowing what the right hand was doing. "Recommendations to eat less of something should not preclude the promotion of a legitimate part of the diet. If the public cannot sort out the two messages — the good taste of butter on the one hand, the moderate use of butter on the other — then we are indeed in trouble."[81]

The butter industry's anti-margarine campaign was aimed at stopping the decline in per-capita butter consumption within three years. Success, coming within eighteen months, surprised the industry greatly. In 1978, butter had possessed its all-time low share of the spread market, 42.7%; by 1980 its share had recovered to 45.4%. The victory was an emotional one, but also had financial implications. The industry regarded the value of an increase of one pound of butter per capita nationally at $85.5 million. The campaign itself cost, in 1980 alone, $6.5 million for all forms of promotion, against an estimated $20 million being spent that year by the "enemy."[82] The war since 1970 had cost the butter side $19.1 million, and margarine $160 million.[83]

For butter, the advertising cost picture escalated to the point that in 1983 $1.57 was spent for every Canadian consumer. In that year, the Ontario Milk Marketing Board spent $12 million for advertising and promotions, up from $10.2 million in 1982. For 1984 the budget rose to $12.9 million. Two thirds of these sums were sent to the Dairy Bureau of Canada for national programs concentrating on butter and cheese. The income was generated by levies on milk sold through the Marketing Board.[84] This campaign had reached the point at which an outside observer could conclude that Ontario dairymen spent more money for advertising than any other farmers in the world. Yet the competition was far ahead in that, in the United States, for every dollar spent on milk advertising, $23 was spent on advertising soft drinks, coffee and teas, while $34 was spent on alcoholic beverage advertising.

Butter's gain came at the expense of margarine. Per-capita annual consumption of margarine had achieved, in 1978, a record 13.2 pounds, and market share stood at an all-time high of 57.3%. At the end of the two-year attack by butter, per-capita consumption had dropped to 12.03 pounds and market share to 54.6%.

The cost of the margarine interests' renewed attack was incredibly high. The number of advertisements increased. Barred from carrying out an explicitly anti-butter campaign because of fear that attacking one of society's "sacred cows" directly would only backfire, the industry intensified the use of the health motif. Fleischmann's again took the lead. Under an ominous heading, "My doctor's advice gave me food for thought," a man in shorts and doing pushups made the following thinly disguised attack on butter.

> My doctor really made me think twice about my health. I'm not taking it for granted any more. So I started exercising. Even fifteen minutes a day makes a big difference. And I started eating better too. You know, the right foods like Fleischmann's margarine. I like the one made with 100% pure corn oil. It has the right levels of polyunsaturates and saturates. That's important when you've been advised to eat carefully. And with Fleischmann's I didn't have to give up great taste just because I started eating right.[85]

Other margarine manufacturers fell in line. Kraft, which once promoted Miracle by stressing its "Country Fresh Flavour,"[86] five years later suggested consumers ought to add Miracle to their "personal plan for good eating."[87]

As the agrarian myth was the dairy industry's main weapon, so the fear of heart disease became that of the margarine industry.

One twist developed: the butter/margarine fight turned into a fight within the family. In June, 1980, Fleischmann's responded to Becel's claim by arguing that (in contrast to Becel's 85% sunflower oil) using 100% pure corn oil was the best way to make margarine.[88] Soon after, Achieve was touted because "[un]like Becel or Fleischmann's, only Achieve is made from 100% sunflower oil," thus reaching a higher polyunsaturate/saturate ratio: 2.4/1 compared to the other two margarines at 2.2/1.[89]

The result of all this competitiveness, with butter and within the margarine industry, has been that the following margarines have led the way since 1984: Parkay with 13% of market share, Imperial 10%, Fleischmann's 9%, Blue Bonnet 9% and Monarch 4%.[90] By 1987 Chefmaster, in third spot, and Lactancia in fifth, had worked their way into the elite ranking.[91]

The margarine forces have recently put several new products on the market: Krona, the margarine that has some animal fat content to make its taste virtually undistinguishable from butter, and Fleischmann's Light, a margarine with half the calories of normal margarines, and a less salty taste.

The effort of the margarine producers to launch their product helped to swing consumers to margarine, and butter's counter-attack redressed the balance. But one cannot avoid the conclusion that there has been a huge waste of effort and money by both sides, especially by the butter industry. In 1982, Jim Romahn, food and agricultural reporter for the *Kitchener-Waterloo Record*, quipped: "Farmers must love to see their very own ads on television. I can think of no other logical explanation for their willingness to shell out millions of dollars, even during these relatively hard times, for fancy television campaigns."[92] There is at present a butter production shortfall across the country because of heavy demand on milk in other product lines. To promote butter sales while butter has to be imported simply does not seem to make sense.[93] However, it is not illogical when one recognizes that the whole range of dairy products is under attack from non-dairy products. For over thirty years, the industry has considered itself in a state of siege.

— *Courtesy Provincial Archives of Alberta, Pollard Collection, P4127, 78.63/1315*

CONCLUSION

Over the past several hundred years there has been a "democratization of dietary gratification."[1] The historian who wrote those words also stipulated that the higher the level of civilization, the more varied the diet becomes.[2] Within the context of the changing patterns of food history of Europe, it is possible to place the Canadian nutritional experience, and even more specifically, the Canadian reaction to margarine's entrance into the food spectrum. Moving from a rural/agricultural to an urban/industrial setting, Canadians reacted to preserve the substance of their way of life.

This aspect of Canadian history occurred within the global context of continuing population growth, with increased pressure on food supplies. Canada has become the producer of large amounts of a select list of foods such as wheat, meats and dairy products. An expanding population, and incorporation into the world pattern of food imports/exports, are the causes of this specialization pattern.

National and world requirements for total food production expansion, as well as the innate desire of people to improve the quality of their life, created a situation in which greater efficiency and closer attention to personal health have become criteria as to which foods are used. As population density increases and pressures on land use build up, Canadians, as well as other members of Western or Atlantic civilization, have been undergoing the switch from a meat and milk diet to one that is more grain-based. The improved efficiency of the cow cannot overcome the greater efficiency of oil seeds such as canola and soy when considered within the expanding demand for food of all sorts. For half of the lifetime of margarine, Canada has not been self-sufficient in dairy products. The position of margarine as a partner with butter as a bread spread enables a greater proportion of milk to be used in its other forms, primarily as fluid milk, which provides the greatest economic return.

The dramatic shift from agriculture to industry provided the context for a major change in lifestyle. The farmer and his family had concentrated upon producing quantity — enough to fill the stomach each day of the year. Aspirations were relatively moderate, and limited by considerations of religion and estate. In the western industrial world, the nineteenth century saw an essential breakdown of these limitations. A higher standard of living, the need for the industrial family to purchase its foods, and a recognition of differing food needs, encouraged the study of nutrition and subsequent action. A well-nourished work force came to be seen as optimal by labour, employer and government.[3] Labour initially rejected margarine because its use was perceived to condone cheap wages. The sounder goal was wages adequate to give a standard of living which would include the better food — butter. Employers accepted margarine for the reasons that labour rejected it — anything to keep costs under control while still providing good nutrition. Government took the position of the Canadian entrepreneur.

Originally butter had powerful advantages over margarine because it was already part of the economic and dietary structure of Canada when margarine first came on the scene. Economic factors have affected the changes in middle-class habits in the last century to a lesser degree than changing tastes, reflecting changes in ideas.[4] Butter, as a food, has served two purposes: to help provide a better diet and, more importantly, to play a role in the transformation of Canadian society from a rural to an urban dominance, steeped in the tradition, prestige and folklore of the agrarian and milk myths as they relate to Canadian culture.[5] Urban Canadians were urged to give cognizance to their rural origins by responding to those roots through the use of butter. Against butter, the prime factor has been geography. Canada cannot maintain self-sufficiency in butter, let alone compete on the world market with countries enjoying a more moderate climate, such as New Zealand and Denmark.

On the other hand, margarine was "born from, through and for" the new industrial age.[6] All margarine oils are interchangeable; there is nothing unique as with butter fat. The invention of margarine was a major factor in the creation of the giant modern agribusiness with its stress on efficiency and profit. Margarine has been a striking example of consumer movement to an alternate food. The industry has centred its battle rationale in the compelling simplicity of the logic in Senator Euler's original contention: an individual ought to have a chance to choose between two foods that serve much the same purpose. The standards of excellence have been those of butter in flavour, consistency, appearance and utility. Margarine has met and even, in the area of cholesterol, surpassed these standards. It is no longer the poor man's substitute for butter, but an alternative used by all classes.

APPENDICES

Appendix 1
Gross Value of Agricultural Production, Quebec and Ontario, 1851-70
($ millions current)

	Ontario			Quebec		
	1851	1860	1870	1851	1860	1870
Wheat	11.0	32.3	14.2	2.7	3.5	2.1
Hay	7.5	15.3	18.0	8.2	12.3	12.3
Oats	3.1	7.9	7.7	2.4	6.5	5.3
Barley	0.3	1.8	6.6	0.2	1.4	1.2
Potatoes	1.6	4.9	5.1	1.4	4.1	5.4
Peas	1.5	3.6	5.4	0.7	1.0	1.5
Other crops	1.7	5.5	6.8	0.9	1.9	2.7
Total crops	26.7	71.3	63.8	16.5	30.7	30.5
Butter and cheese	2.6	4.9	8.4	1.5	2.8	4.5
Wool	0.7	1.1	1.8	0.4	0.6	0.8
Livestock	6.0	11.9	19.8	3.1	5.8	10.6
Total	36.0	89.2	93.8	21.5	39.9	46.4

Source: McCallum, *Unequal Beginnings*, 127.

Appendix 2
Butter Production in Canada, 1871-91
(in pounds)

	1871		1881		1891	
	Pounds	%	Pounds	%	Pounds	%
Canada	74,190,583	100	102,534,466	100	111,577,210	100
N.S.	7,161,867	9.7	7,465,285	7.4	9,011,118	8.1
N.B.	5,115,947	6.9	6,517,176	6.4	7,798,268	7.0
Quebec	24,289,127	32.7	30,630,397	29.9	30,113,226	27.0
Ontario	37,623,642	50.7	54,862,365	53.5	55,564,496	49.7
Manitoba	—		957,152	0.9	4,830,368	4.3
P.E.I.	—		1,688,190	1.6	1,969,213	1.8
B.C.	—		343,387	0.3	393,089	0.4
Terr.	—		70,717	0.06	1,897,432	1.7

Source: *Census of Canada*, 1871, 1881, 1891.

Appendix 3
Participants in Margarine Debate

Name	Party	Occupation	Riding[a]	Riding Population	Home Butter Production	Factories/ Value
B. Allen	Lib	Merchant	N. Grey	23,334	860,003	—
J. Armstrong	Lib	Warden	S. Middlesex	18,888	—	—
T. Bain	Lib	Farmer	N. Wentworth	15,998	565,878	—
D. Bergin	Con	Doctor	Cornwall & Stormont	23,198	1,307,380	—
E. Blake	Lib	Lawyer	W. Durham	17,555	526,544	—
M. Bowell	Con	Dairy	N. Hastings	20,479	512,582	—
J. Carling	Con	Business	London	19,746	1,930	—
G. Casey	Lib	Lawyer	W. Elgin	23,480	442,318	—
E. Cochrane	Con	Farmer	E. Northumberland	22,991	684,274	—
J. Costigan	Con	Politic	Victoria, N.B.	15,686	214,634	—
C. Ferguson	Con	Doctor	N. Leeds & Grenville	12,423	840,000	1/$7000
S. Fisher	Lib	Farmer	Brome, P.Q.	15,827	1,598,350	—
A. Gillmor	Lib	Business	Charlotte, N.B.	26,087	614,295	—
G. Guillet	Con	Merchant	W. Northumberland	16,984	423,275	—
S. Hesson	Con	Merchant	N. Perth	26,538	857,371	—
C. Hickey	Con	Doctor	Dundas	20,598	1,843,866	—
J. Jenkins	Con	Doctor	Queens, P.E.I.	48,111	832,270	—
G. Lankerkin	Lib	Doctor	S. Grey	25,703	1,003,947	1/$2300
D. Mills	Lib	School Sup.	Bothwell	22,477	730,222	—
A. McLelan	Con	Business	Colchester, N.S.	26,720	631,267	—
J. McMullen	Lib	Merchant	N. Wellington	26,024	909,212	1/$1500
G. Orton	Con	Doctor	C. Wellington	26,816	781,004	—
W. Paterson	Lib	Business	S. Brant	20,482	436,425	—
J. Platt	Lib	Doctor	Prince Edward	21,044	683,131	—
T. Sproule	Con	Doctor	E. Grey	25,334	1,221,384	—
G. Taylor	Con	Merchant	S. Leeds	20,716	836,353	3/$45,350
J. White	Con	Business	E. Hastings	17,400	461,027	—
J. Wood	Con	Lawyer	Brockville	15,107	169,669	—

a. Ontario, unless stated otherwise.
Source: *Census of Canada*, 1880-81, and Gimmell, *The Canadian Parliamentary Companion 1885* (1885).

Appendix 4
Milk Cows in Canada, 1871-1921

	1871			1881			1891			1901			1911			1921		
	No.	%		No.	%	Decade % Incr.	No.	%	Decade % Incr.	No.	%	Decade % Incr.	No.	%	Decade % Incr.	No.	%	Decade % Incr.
Canada	1,251,209	100		1,595,800	100	+21.60	1,857,112	100	+14.07	2,408,677	100	+22.90	2,595,255	100	+7.75	2,744,700	100	+5.8
Ontario	638,759	51.06		782,243	49.01	+18.34	876,167	47.18	+10.72	1,065,763	44.25	+17.79	1,032,976	39.80	+3.07	973,308	35.46	-5.8
Quebec	406,542	31.49		490,977	30.77	+17.20	549,544	29.59	+10.66	767,825	31.88	+28.43	754,220	29.06	+1.77	796,029	29.00	+5.5
N.S.	122,688	9.80		137,639	8.63	+10.86	141,684	7.63	+2.85	138,817	5.76	-2.20	129,274	4.98	+6.87	121,460	4.42	-6.04
N.B.	83,220	6.65		103,965	6.51	+19.95	106,649	5.74	+2.52	111,084	4.61	+4.00	108,557	4.18	+2.27	108,655	3.95	+0.1
Manitoba				20,355	1.27		82,712	1.41	+75.39	141,481	2.35	+41.54	155,328	6.98	+9.78	167,498	6.10	+7.8
B.C.				10,878	0.68		17,504	0.94	+37.85	24,535	1.03	+28.66	33,954	1.31	+38.39	45,245	1.65	+33.3
P.E.I.				45,895	2.88		45,849	2.47	+0.10	56,437	2.34	+18.76	52,109	2.01	+7.66	46,666	1.70	10.4
Territories				3,848	0.24		37,003	1.99	+89.60	102,735	4.27	+63.92						
Alberta													147,649	5.69		218,353	7.95	+47.9
Sask.													181,168	6.98		267,486	9.75	+47.6

Source: *Census of Canada*, 1871, 1881, 1891, 1901, 1911, 1921.

APPENDICES 169

Appendix 5
Butter Imports into the United Kingdom from British Trade Returns

	1905	1906	1907	1908	1909	1910	1911	1912
Cwt. imported from:								
Russia	461,140	606,549	657,649	639,118	601,712	584,040	638,284	683,650
Sweden	188,209	182,803	226,740	238,929	312,142	345,684	360,357	335,014
Denmark	1,630,363	1,675,761	1,818,811	1,857,103	1,764,027	1,726,091	1,707,178	1,618,048
Germany	5,372	10,701	7,297	3,195	2,965	3,481	–	–
Netherlands	209,897	195,366	168,496	244,356	148,567	154,537	104,655	113,716
France	348,422	319,401	281,306	394,612	413,306	361,249	171,080	246,652
United States	84,874	157,312	1,063	39,540	693	756	23,052	2,596
Australia	450,293	545,827	598,986	409,106	384,619	639,093	874,399	541,253
New Zealand	300,418	311,672	313,863	221,395	278,581	362,674	276,446	349,012
Canada	292,117	190,968	34,753	47,877	22,522	16,805	61,936	27
Other countries	176,741	140,898	101,192	115,590	133,699	131,129	85,305	115,191
Total	4,147,866	4,337,258	4,210,156	4,210,821	4,062,833	4,325,539	4,302,692	4,005,159
Percentage imported from:								
Russia	11.1	14.1	15.6	15.2	14.8	13.5	14.8	17.1
Sweden	4.5	4.2	5.4	5.7	7.7	7.9	8.4	8.4
Denmark	39.3	38.6	43.2	44.1	43.4	39.9	39.7	40.4
Germany	0.1	0.3	0.2	0.1	0.07	0.09	–	–
Netherlands	5.1	4.5	4.0	5.8	3.7	3.9	2.4	2.8
France	8.4	7.1	6.7	9.7	10.1	8.4	4.0	6.1
United States	2.0	3.6	0.03	0.9	0.01	0.01	0.5	0.0
Australia	10.8	12.5	14.2	9.5	9.5	14.7	20.3	13.6
New Zealand	7.5	7.2	7.5	5.3	6.9	8.3	6.4	8.7
Canada	7.0	4.4	0.8	1.1	0.6	0.3	1.4	0.0
Other countries	4.2	3.5	2.4	2.6	3.3	3.0	2.1	2.9

Appendix 5 – Continued
Cheese Imports into the United Kingdom from British Trade Returns

	1905	1906	1907	1908	1909	1910	1911	1912
Cwt. imported from:								
Canada	1,858,767	1,925,835	1,698,847	1,541,502	1,566,546	1,607,064	1,473,275	1,352,570
United States	175,256	233,445	114,300	105,555	54,617	38,427	150,321	21,227
Netherlands	214,033	229,341	241,551	279,401	285,329	231,832	207,917	28,286
New Zealand	78,626	126,216	192,301	264,995	368,531	453,785	397,845	543,917
Other countries	116,000	123,957	125,234	114,633	115,067	125,427	118,964	122,799
Total	2,422,682	2,638,794	2,372,233	2,306,086	2,390,090	2,456,340	2,348,326	2,308,799
Percentage imported from:								
Canada	76.1	73.0	71.7	66.8	65.5	65.5	62.7	58.6
United States	7.2	8.8	4.8	4.6	2.3	1.6	6.4	0.9
Netherlands	8.7	8.7	10.2	12.1	12.0	9.4	8.9	11.6
New Zealand	3.2	4.8	8.1	11.5	15.4	18.5	16.9	23.6
Other countries	4.8	4.7	5.2	5.0	4.8	5.0	5.1	5.3

Source: Canada, "Report of the Dairy and Cold Storage Commissioner... 1913," *Sessional Papers, 1914*, No. 15a, 132.

Appendix 6
Butter and Margarine Production and Consumption
1909-10 to 1922-23
(000's pounds)

Year	Creamery Butter		Dairy Butter		Total Butter Production	Butter Imports	Butter Exports	Total Domestic Butter Consumption	Margarine: Manufactured and Imported	Total Consumption Butter and Margarine
1909-10	64,698	(32%)	137,110	(68%)	201,868	393	4,615	197,586	—	197,586
1910-11	n/a[a]		n/a		n/a		1,227	3,143	n/a	—
1911-12	n/a		n/a			3,987	8,894	n/a	—	n/a
1912-13	n/a		n/a			7,989	828	n/a	—	n/a
1913-14	n/a		n/a			7,317	1,228	n/a	—	n/a
1914-15	83,991		n/a		n/a	6,822	2,724	n/a	—	n/a
1915-16	82,564		n/a		n/a	4,309	3,441	n/a	—	n/a
1916-17	87,526		n/a		n/a	997	7,990	186,300	—	186,300
1917-18	93,298	(48%)	100,000	(52%)	193,300	434	4,926	188,800	—	188,800
1918-19	103,899	(51%)	100,000	(49%)	203,900	1,939	13,659	192,200	16,963[b]	209,163
1919-20	122,776	(57%)	103,487	(43%)	215,100	397	17,612	197,900	12,948	210,848
1920-21	128,745	(54%)	107,379	(46%)	236,124	3,741	9,739	230,126	10,855	219,271
1921-22	152,502	(58%)	107,222	(42%)	259,724	6,079	8,430	257,373	3,242	260,615
1922-23	162,835	(60%)	110,610	(40%)	273,445	3,768	21,994	255,219	3,287	258,505
1923-24	178,894	(62%)	108,535	(38%)	287,429	1,558	13,649	299,520	2,626 (9 mos)	302,146

a. Not available.
b. Source noted total for 16 months from 1 December 1917 to 21 March 1919.

Source: *Census of Canada*, 1911 and 1921; various *Canada Year Books*; *Historical Statistics* (1965), series 253 and 254; and *Reference to ... Section 5A of Dairy Industry Act*, R.S.C. 1927, cap. 45, 6.

Appendix 7
Butter and Margarine:
Pounds Manufactured and Imported
1917-23

Oleomargarine	Manufactured lbs.	Imported lbs.	Total lbs.
1 Dec. 1917 to 31 Mar. 1919	10,183,179	6,480,130	16,963,609
Year ended 31 Mar. 1920	6,450,902	6,497,031	12,917,933
Year ended 31 Mar. 1921	6,224,422	4,660,747	10,855,169
Year ended 31 Mar. 1922	1,902,629	1,339,748	3,212,377
Year ended 31 Mar. 1923	2,422,029	1,165,449	3,287,169
6 months ended Sept. 1923	1,880,678[a]	745,015	2,625,693
Total	31,063,839	20,858,411	51,922,250

a. "Manufactured" covers five months ended August 1923.

Butter	Manufactured	Imported	Total
1918	193,300,000[a]	400,000	193,700,000
1919	203,900,000[a]	1,900,000	205,800,000
1920	215,100,000	400,000	215,500,000
1921	228,700,000	3,700,000	232,400,000
1922	252,500,000	6,000,000	258,500,000
1923	262,800,000	3,700,000	266,500,000
Total	1,356,300,000	16,100,000	1,372,400,000

a. Includes estimated dairy butter production of approximately 100 million pounds per year. Statistics on dairy butter production are not available for the years previous to 1920.
Source: "In the Matter of a Reference as to the Validity of Section 5(a) of the Dairy Industry Act," R.S.C., 1927 cap. 45, 6.

Margarine Imports, 1918-24
(pounds and dollar values)

	1918	1919	1920	1921	1922	1923	1924
United Kingdom							
Lbs.					6,000		
$					1,399		
United States							
Lbs.	2,262,514	4,217,916	6,497,031	4,630,747	1,339,784	1,165,440	745,015
$	607,302	1,180,656	1,872,204	1,206,351	255,994	190,782	130,605
Total Imports for Consumption							
Lbs.	2,262,514	4,217,916	6,497,031	4,630,747	1,345,784	1,165,440	745,015
$	607,302	1,180,656	1,872,104	1,206,351	257,393	190,782	130,605

Source: Canada, *Sessional Papers, 1922*, No. 10b, *Trade of Canada*, 467, and *Annual Departmental Reports, 1924-25*, IV, *Trade of Canada*, 439.

Appendix 8
National Standard for Margarine

B.09.016. [S]. Margarine

(a) shall be a plastic or fluid emulsion of water in fats, oil or fats and oil that are not derived from milk and may have been subjected to hydrogenation;
(b) shall contain
    (i)   not less than 80 per cent fat, oil or fat and oil calculated as fat, and
    (ii)   notwithstanding section D.01.009, not less than
        (A)   3300 International Units of vitamin A, and
        (B)   530 International Units of vitamin D per 100 grams; and
(c) may contain
    (i)   skim milk powder or butter milk powder,
    (ii)   whey solids or modified whey solids,
    (iii)   protein,
    (iv)   water,
    (v)   vitamin E, if added in an amount that results in not less than 0.6 International Units of alphatocopherol per gram of linoleic acid present in the margarine,
    (vi)   beta-carotene, annatto, turmeric, canthaxanthin, beta-apo-8' carotenal and ethyl beta-apo-8' carotenoate as colouring, but beta-apo-8' carotenal and ethyl beta-apo-8' carotenoate shall not be added, singly or in combination, in an amount exceeding 35 parts per million,
    (vii)   a flavouring agent,
    (vii)   a sweetening agent,
    (ix)   sodium chloride and potassium chloride,
    (x)   monoglycerides, mono and diglycerides, or any combination thereof, in an amount not exceeding 0.5 per cent,
    (xi)   lecithin in an amount not exceeding 0.2 per cent,
    (xii)   sorbitan tristearate in an amount not exceeding 1.0 per cent,
    (xiii)   sorbic acid and benzoic acid and their salts either singly or in any combination in an amount not exceeding 1000 parts per million expressed as the acids,
    (xiv)   butylated hydroxyanisole, butylated hydroxytoluene and propyl gallate either singly or in any combination in an amount not exceeding 0.01 per cent of the fat content,
    (xv)   ascorbyl palmitate and ascorbyl stearate either singly or in combination in an amount not exceeding 0.02 per cent of the fat content,
    (xvi)   monoglyceride citrate, monoisopropyl citrate and stearyl citrate either singly or in any combination in an amount not exceeding 0.01 per cent of the fat content,
    (xvii)   citric and lactic acids and their potassium and sodium salt,
    (xviii)   tartaric acid and sodium potassium tartrate,
    (xix)   sodium or potassium bicarbonate, sodium or potassium carbonate and sodium or potassium hydroxide, and
    (xx)   calcium disodium ethylenediamine tetraacetate in an amount not exceeding 75 parts per million.

B.09.017. [S]. Calorie-Reduced Margarine

(a) shall conform to the standard for margarine except it shall contain not more than
  (i) 40 per cent fat, oil or fat and oil calculated as fat, and
  (ii) 50 per cent of the calories that would be normally present in the product if it were not calorie-reduced;
(b) may contain
  (i) acacia gum,
  (ii) agar,
  (iii) algin,
  (iv) carob bean gum,
  (v) carrageenan,
  (vi) furcelleran,
  (vii) guar gum,
  (vii) karaya gum,
  (ix) propylene glycol alginate,
  (x) tragacanth gum, or
  (xi) xanthan gum,
  either singly or in any combination, in an amount not exceeding 0.5 per cent; and
(c) notwithstanding subparagraph B.09.016(c)(xi), may contain lecithin in an amount not exceeding 0.5 per cent.

B.09.020

Subject to sections B.09.021, B.24.016, B.24.103 and B.24.202, no person shall, in an advertisement to the general public or on a label of a food, make any representation, general or specific, respecting the fatty acid or cholesterol content of the food.

B.09.021

Where the proportion of cis-methylene interrupted polyunsaturated fatty acids contained in the total fat is at least 40 per cent in the case of an oil and at least 25 per cent in the case of shortening, margarine or a margarine-like product, and the proportion of saturated fatty acids does not exceed 25 per cent in either case, a person may, in an advertisement or on the label of such product state:
(a) the percentage of saturated fatty acids, on a total fat basis; and
(b) the percentage of polyunsaturated fatty acids on a total fat basis, where it is calculated as total cis-methylene interrupted polyunsaturated fatty acids,

if those statements are grouped together and given equal prominence in the advertisement or on the label.

B.09.022

No person [sic] sell cooking oil, margarine, salad oil, simulated dairy product, shortening or food that resembles margarine or shortening, if the product contains more than 5 per cent C22 Monoenoic Fatty Acids calculated as a proportion of the total fatty acids contained in the product.

NOTES

Introduction

1 See Charles Wilson, *History of Unilever: A Study in Economic Growth and Social Change* (New York: Praeger, 1968), vol. II, chapter IV.
2 Katharine Snodgrass notes that the first Canadian legislation prohibiting manufacture, sale or importation was passed in 1914, rather than 1886 (*Margarine as a Butter Substitute* [Stanford: Stanford University Press, 1930], 117).
3 Ibid., 125-29; A.E. Bailey, *Bailey's Industrial Oil and Fat Products* (New York: John Wiley, 1964), 255-57.
4 The *Reader's Guide to Periodical Literature*, XIII, 1941-1943 was the first to have a reference to margarine, with oleomargarine remaining the prime reference; XXI, 1957-1959 first used margarine as the chief entry. In Canadian law, margarine, as distinct from oleomargarine, was first mentioned in the Dairy Act of 1914.
5 Snodgrass, *Margarine*, 120.
6 Canada, Senate, *Debates, 1946*, 83-84.

Chapter 1

1 William L. Marr and Donald G. Paterson, *Canada: An Economic History* (Toronto: Macmillan, 1980), 186, table 6:12.
2 John McCallum, *Unequal Beginnings: Agricultural and Economic Development in Quebec and Ontario until 1870* (Toronto: University of Toronto Press, 1980), 50.
3 M.C. Urquhart and K.A.H. Buckley (eds.), *Historical Statistics of Canada* (Cambridge: Cambridge University Press, 1965), series C 1-7.
4 McCallum, *Unequal Beginnings*, 3.
5 Roy E. George, *A Leader and a Laggard: Manufacturing Industry in Nova Scotia, Quebec and Ontario* (Toronto: University of Toronto Press, 1970), 10, 15, and 22. S.A. Saunders, *The Economic History of the Maritime Provinces*, Study prepared for the Royal Commission on Dominion-Provincial Relations (Ottawa, 1939), 62.
6 Trevor H. Levere and Richard A. Jarrel (eds.), *A Curious Fieldbook: Science and Society in Canadian History* (Toronto: Oxford University Press, 1974), 160-61.
7 Ibid., 162.
8 Marvin McInnis, "The Changing Structure of Canadian Agriculture, 1867-1897," *Journal of Economic History*, 42, 1 (March 1982): 193 and 195.
9 McCallum, *Unequal Beginnings*, 52.
10 Ibid., 5.
11 *Census of Canada, 1881*, III, 234 ff., 248, 483.
12 Quebec, "General Report of the Commissioner of Agriculture and Public Works for 1880/81," *Sessional Papers, 1883*, 417-18.
13 Ibid.
14 Canada, House of Commons, *Journals, 1884*, Appendix No. 6, 2-3.
15 Catherine Parr Traill, *Canadian Settler's Guide*, New Canadian Library #4 (Toronto: McClelland and Stewart, 1969), 188.
16 Quebec, "General Report of the Commissioner of Agriculture and Public Works for 1880/81," *Sessional Papers, 1883*, No. 2, 420.
17 Edwin C. Guillet, *Pioneer Farmer and Backwoodsman* (Toronto: Ontario Publishing Company, 1963), 2:87.
18 J.A. Ruddick, *Dairying Industry in Canada* (Ottawa: Dept. of Agriculture, 1911), 37-39.
19 Marr and Paterson, *Canada: An Economic History*, 104.
20 Ian Drummond, *Progress Without Planning: The Economic History of Ontario from Confederation to the Second World War* (Toronto: University of Toronto Press, 1987), 129.
21 McInnis, "Changing Structures of Canadian Agriculture," 193.

22 Commons, *Journals, 1884*, Appendix No. 6, a paper presented by E.A. Bainard, Director of Agriculture in the Quebec Department of Agriculture, 38.
23 Ontario, Ontario Agricultural Commission, *Report, 1881*, 405.
24 Ibid., p. 418, stipulated that two-thirds of Canadian butter was "best shipping brand." On the other hand E.A. Bainard of the Quebec Department of Agriculture listed finest at 5%-10%, fine 25%-30% and poor 50%-60%. Commons, *Journals, 1884*, Appendix No. 6, 31.
25 Robert L. Jones, *History of Agriculture in Ontario: 1613-1880* (Toronto: University of Toronto Press, 1946), 260.
26 Commons, *Journals, 1884*, Appendix No. 6, 20, 23, 39.
27 Ontario Agricultural Commission, *Report, 1881*, 415.
28 Robert E. Ankli and Wendy Millar, "Ontario Agriculture in Transition: The Switch from Wheat to Cheese," *Journal of Economic History*, 42, 1 (March 1982): 207-15.
29 Jones, *History of Agriculture in Ontario*, 260.
30 Commons, *Journals, 1884*, Appendix No. 6, 30.
31 Ibid. and Ontario Agricultural Commission, *Report, 1881*, Appendix 26.
32 Canada, House of Commons, "Report of the Minister of Agriculture, 1877," *Sessional Papers, 1878*, No. 9, vi-vii. The 1891 report is the first to give a separate subsection to dairying.
33 Canada, House of Commons, *Debates* on film for May 19, 1874. Canada, *Statutes of Canada, 1874*, 37 Vic, cap. 8, 39-46. "An Act to impose License duties on Compounders of Spirits; to amend the Act respecting the Inland Revenue; and to prevent the adulteration of Food, Drink and Drugs."
34 Commons, *Debates, 1878*, 2032-33. *Statutes of Canada*, 1878, 40 Vic, cap. 11, 72-3. "An Act to amend the Act thirty-seventh Victoria, chapter eight, entitled, 'An Act to Impose license duties on compounders of spirits; to amend the Act respecting the Inland Revenue, and to prevent the Adulteration of Food, Drink and Drugs.'"
35 "Oleomargarine," *Farmer's Advocate*, September 1878, 207.
36 Ibid., April 1880, 77.
37 Ibid., May 1884, 138-39.
38 Ibid., July 1884, 190.
39 Ontario Archives, *Analysis of Reports of the Agricultural and Horticultural Societies, 1882*, 26.
40 Commons, *Journals, 1884*, Appendix No. 6, 2-3.
41 "Dairymen's Association," *Morning Chronicle* (Halifax), 26 March 1886, 1.
42 Commons, *Debates, 1886*, 765.
43 Ontario, Legislative Assembly, "Forty-first Annual Report of the Agricultural and Arts Association of Ontario," *Sessional Papers, 1887*, No. 5, 10.
44 Commons, *Debates, 1886*, 554.
45 Ontario, "Annual Reports of the Dairymen and Creameries' Associations of the Province of Ontario, 1894," *Sessional Papers, 1895*, No. 21, 7.
46 Commons, *Debates, 1886*, 547.
47 Ibid., 554.
48 Commons, *Journals, 1886*, 86 and 108.
49 *Statutes of Canada, 1867*, 31 Vic, cap 7, 147.
50 Commons, *Debates, 1886*, 487, 549, 601.
51 Ibid., 759. See also William Paterson's statements later in the House, ibid., 1188-89.
52 Of the 191 ridings in Canada, 54 (28%) had significant butter production of at least 700,000 pounds annually, according to the 1881 census. Of these, 54 ridings 14 (26%), had their Member speak in the margarine debate of 1886. Looking at it another way, of the 26 backbenchers involved in the debate, 15 (58%) came from ridings with significant butter production.
53 Samuel P. Hays, *Response To Industrialism, 1885-1914* (Chicago: University of Chicago Press, 1957), and Robert Wiebe, *Search For Order, 1877-1920* (New York: Hill and Wang, 1967).
54 Commons, *Debates, 1886*, 547.
55 Ibid., 487.
56 Ibid., 1338.

57 "The great centre at which oleomargarine is prepared is Hunter's Point, in the State of New York. It is true, as a gentleman behind me remarks, that there are large factories at Chicago also, but the manufactories at Hunter's Point are the largest in the States. At that point the factories are entirely in the hands of Jews. Dead cattle, dead hogs, dead horses, dead cats, too, I might say, are used for the production of this article, which is thrown upon the market as human food and as a substitute for one of the most wholesome and healthful foods employed by the people. Animals which die in the cars from overcrowding, or are strangled to death, animals that die from disease, are purchased at a nominal sum by these people. Hogs that die from disease, from hog cholera, and from charbon [*sic*], horses that die, and cattle that die from pleuro-pneumonia are used in the manufacture; and the counties around New York furnish a large number of cattle which die in the distilleries and in the cattle byres from that terribly infectious disease, pleuro-pneumonia, and are taken to these factories by the hundred and by the thousand, many of them in such a state that if you take hold of them by the leg you draw the leg off. These are thrown into vats where they are subjected to a heat of over 300 degrees, until the fat separates from the flesh and the bones and hides and floats upon the top, and the offal goes to the bottom. When this separation takes place, after the heat has been applied long enough — a heat that, I make bold to say, does not destroy the germs of disease in those dead carcasses — the vacuum pump is applied, the vats are cooled until the temperature comes down to about 150 degrees, when ice is thrown in and after a time the fat separates and becomes hard. Then the manholes are opened, water is poured in until a couple of feet of water are between the offal and the hardened fat, the water is run off, the offal is allowed to escape from the bottom, the fat is taken out and then bleached by the application of sulphuric acid and other chemicals. It is contended by the parties who own those factories and make human food out of the rotting, decayed carcasses that the chemicals destroy all the diseased particles. No proof of that has ever been produced. On the contrary, we know, as was stated by the hon. member for Dundas (Mr. Hickey) the other night, that where animals which died of a certain disease, were buried as deep as eight or ten feet under ground, a year afterwards healthy animals feeding on the ground under which those diseased animals had been buried were infected with the same disease and it spread among the cattle throughout the surrounding country. Yet this is the material from which human food, an article designed to take the place of butter, is manufactured, and it can be produced at so cheap a rate that it is impossible for the honest dairyman or for the honest butter-maker to compete. In preparing this food the manufacturers designedly go to work to deceive. By machinery, after they have bleached this fat to the color they desire, by means of chemicals so as to place the article on the market, they granulate it so as to give it the appearance of butter. They imitate what is called gilt-edge butter so closely that it is impossible for anyone to tell whether it is the genuine article or not. The manufacturers also add chemicals so as to change the odor to that of first-class butter, and they add chemicals to give it a flavor. Whether those chemicals added to make oleo-margarine are likely to add to the healthfulness of the product or not, I leave for the House to declare" (ibid., 686).

58 Ibid., 686–88.
59 Ibid., 549–53.
60 Ibid., 547 ff.
61 Ibid., 549. An analysis of the results in Fisher's riding of Brome of the 1882 and 1887 general elections shows that Fisher did not suffer from his initial pro-margarine stance. His back-pedalling, even as early as the late stages of this 1886 debate, may help to explain his success.
62 Ibid., 758–69.
63 Ibid., 758–68.
64 Ibid., 1205.
65 Ibid., 548.
66 Ibid., 550–52.
67 Ibid., 758.

68 "Dominion Parliament," *Morning Chronicle* (Halifax), 8 April 1886, 3; ibid., 19 April 1886, 3; "Ottawa Letter," ibid., 26 April 1886, 1; and "Dominion Parliament," *Daily Times* (Moncton), 2 June 1886, 3.
69 "Echoe d'Ottawa," 24 May 1886, 4.
70 10 June 1886, 2. The whole series of articles ran in the *Journal* from 11 March to 12 August 1886.
71 "The Tories and Oleomargarine" (editorial), 17 April 1886, 4.
72 "Things to Do and to Know," 12 February 1886, 3.
73 14 May 1886, 1293.
74 "Butter Substitute," 17 April 1886, 10.
75 "Counterfeit Butter," 9 April 1886, 4; "Oleomargaine," 12 April 1886, 4; "Counterfeit Butter Scandal," 9 April 1886, 4; and "And Now Exclude Butter," 13 May 1886, 4.
76 18 May 1886, 3.
77 Commons, *Debates, 1886*, 758-69.
78 Commons, *Journals, 1886*, 159-60.
79 Ibid., 163.
80 Commons, *Debates, 1886*, 1188.
81 Ibid., 1188-95. See Paterson's statement for a development of the confused situation.
82 Ibid., 1204.
83 Commons, *Journals, 1886*, 319, 350, 352.
84 Senate, *Debates, 1886*, 920-21.
85 Commons, *Debates, 1886*, 1777; Senate, *Debates, 1886*, 920-21; and "An Act to Prohibit the Manufacture and Sale of Certain Substitutes for Butter," *Statutes of Canada, 1886*, 227, 49 Vic, cap. 42.
86 Commons, *Debates, 1886*, 1204, and *Statutes of Canada, 1886*, 221, 49 Vic, cap. 37.
87 James Harvey Young, "This Greasy Counterfeit: Butter Versus Oleomargarine in the United States Congress, 1886," *Bulletin of the History of Medicine*, 53, 1 (1979): 392-414.

Chapter 2

1 *Farmer's Advocate*, 15 February 1894, 63.
2 Since numbers of producers and size of each producer's herd are not available for this nineteenth-century era, size of provincial and national herds is the best evidence available.
3 Canada, *Census, 1921, V*, cv, Table civ, "Production and Value of Milk by Provinces, 1920 and 1910."
4 *Canada Year Book, 1911*, xxv; *1921*, 250.
5 Provincial Archives of Alberta (hereafter PAA), Alberta, Department of Agriculture, *Annual Report, 1907*, 6 and 78.
6 Canada, "Annual Report of Minister of Agriculture for 1893," *Sessional Papers, 1894*, No. 8, xxvi.
7 Ibid., xx.
8 Between 1884 and 1897, butter was eight times in the top-ten list of foods suffering from the most adulteration. See Craig Robert Stuart, "Genesis of Federal Food and Drug Legislation in Canada, 1874-1890," Master's cognate essay, Wilfrid Laurier University, 1985.
9 Canada, "Annual Report of Minister of Agriculture for 1893," *Sessional Papers, 1894*, No. 8, viii-ix, and "Report, 1895 for 1894," xviii.
10 Ibid., *1895*, No. 8, 1894, xix.
11 Alberta, Department of Agriculture, *Annual Report, 1907*, 78.
12 Earl Allan Haslett, "Factors in the Growth and Decline of the Cheese Industry in Ontario, 1864-1924," Ph.D. dissertation, University of Toronto, 1969, 111.
13 Ontario, "Annual Report of Minister of Agriculture, 1894," *Sessional Papers, 1895*, No. 21, 46.
14 Canada, "Report, 1895," *Sessional Papers, 1896*, No. 8, xvii – xix.

15 Ibid., "Report, 1890," *1891*, No. 6, xiv.
16 Ibid., "Report, 1891," *1892*, No. 7, xii, and *1893*, for 1892, xviii.
17 Ibid., "Report, year ending March 31, 1914," *1915*, No. 15, 13.
18 Ibid., "Report, 1892," *1893*, No. 7, xviii; Alberta, Department of Agriculture, *Annual Report, 1907*, 74; and Gordon C. Church, *An Unfailing Faith: A History of the Saskatchewan Dairy Industry*, microform (Toronto: Micromedia, 1985), 38.
19 *Historical Statistics* (1983), series M425.
20 Canada, "Report of the Dairy and Cold Storage Commissioner for the Year Ending March 31, 1914," *Sessional Papers, 1915*, No. 15a, 3.
21 Marr and Paterson, *Canada: An Economic History*, 151, 153.
22 For a discussion of this point see ibid., 184-87.
23 Canada, "Report of the Dairy and Cold Storage Commissioner for Year Ending March 31, 1914," *Sessional Papers, 1915*, No. 15a, 2, 13.
24 Ibid., "Annual Report of Minister of Agriculture for Year Ending March 30, 1913," *1914*, No. 15, 11.
25 Reay Tannahill, *Food in History* (London: Eyre Methuen, 1973), 375-77.
26 *Farmer's Advocate*, November 1901.
27 Ibid., 15 November 1901, 741.
28 "Butter Fat as a Food" (editorial), *Maritime Farmer and Cooperative Dairyman*, 10 July 1917, 550.
29 Tannahill, *Food in History*, 362.
30 *Farmer's Advocate*, 4 December 1913, 2100.
31 Ibid., 25 December 1913, 2289.
32 Ibid., 23 April 1914, 807.
33 Commons, "Report of the Select Standing Committee on Agriculture and Colonization, 1898," *Journals, 1898*, Appendix 3, 56; evidence by James W. Robertson, Commissioner of Agriculture and Dairying for Canada.
34 Process butter was a product derived from the rechurning of butter with either buttermilk or skim milk. The only complaint about it, other than the possibility of confusion with real butter, was that it became rancid more readily than butter.
35 Renovated butter was butter that had been rechurned for the purpose of increasing the water content.
36 Commons, *Debates, 1903*, 5052-56.
37 Ibid., 5052-84, 5419-45. "An Act to Prohibit the Importation, Manufacture or Sale of Adulterated, Process or Renovated Butter, Oleomargarine, Butterine or Other Substitute for Butter, and to Prevent the Improper Marking of Butter," *Statutes of Canada, 1903*, 3 Edward VII cap 6.
38 Ibid., *1908*, 10554.
39 Ibid., *1911*, 6504 and 6668.
40 Ibid., *1914*, 1751-52, 2384-88 and 3162-73.
41 Senate, *Debates, 1914*, 433-39, and "An Act to Regulate the Manufacture and Sale of Dairy Products and to Prohibit the Manufacture or Sale of Butter Substitutes," *Statutes of Canada, 1914*, 4-5 Geo. V, cap. 7.
42 *Farmer's Advocate*, 2 April 1914, 637.

Chapter 3

1 *Halifax Herald*, 30 April 1917, 29.
2 "The Case Against Oleo" (editorial), *Farmer's Advocate*, 9 November 1916, 1835.
3 Saskatchewan, Department of Agriculture, *Annual Report 1916 of Dairy Commissioner*, 32.
4 Ibid., W.A. Wilson, Dairy Commissioner, Regina, to Hon. Martin Burrell, Minister of Agriculture, Ottawa, 6 December 1916.

5 Commons, "Annual Report of the Minister of Agriculture for Year Ending March 31, 1917," *Sessional Papers, 1918*, No. 15, 10.
6 "Butter," Bulletin No. 373, "Reports, Returns and Statistics of the Inland Revenues, 1917-1918," *Sessional Papers, 1918*, No. 14, 6-7.
7 Robert Craig Brown and Ramsay Cook, *Canada, 1896-1921: A Nation Transformed*, (Toronto: McClelland and Stewart, 1974), chap. 11.
8 Saskatchewan, Department of Agriculture, Dairy Branch, "Oleomargarine, 1916-1948," W.A. Wilson to J.A. McFeeter, 22 January 1917.
9 Ibid., Neil S. Dow, Assistant Manager, De Laval Dairy Supply Co., Winnipeg to W.A. Wilson, Dairy Commissioner, Regina, 18 January 1917; T.J. Coyle, commission broker, Winnipeg to W.A. Wilson, 18 January 1917; W.A. Wilson to J.G. Turiff, MP, Ottawa, 25 January 1917; and J.A. McFeeter, Secretary-Treasurer, Toronto Creamery Co., Toronto to W.A. Wilson, 30 January 1917.
10 Commons, *Debates, 1918*, 292.
11 "Oleomargarine to British Columbia is Viewed With Alarm," *Victoria Daily Times*, 24 October 1917, 15.
12 "Oleomargarine Sale Protected," *Winnipeg Free Press*, 23 January 1917, 13.
13 PAA, United Farmers of Alberta, *Annual Report and Yearbook* (1917), 199.
14 Saskatchewan Archives Board (hereafter SAB), Saskatchewan Grain Growers' Association, *Minutes and Reports*, 1. Convention *Minutes*, 13-16 February 1917, 63.
15 "Manitoba Dairymen Annual Convention," *Winnipeg Free Press*, 16 February 1917, 15; Saskatchewan, Department of Agriculture, Dairy Branch, "Oleomargarine, 1916-1948," Wilson to McFeeter, 22 January 1917; and ibid., G. Marker to W.A. Wilson, 17 February 1917.
16 Ibid., J.A. McFeeter, Secretary-Treasurer, Toronto Creamery Co., Toronto to W.A. Wilson, 23 November 1916; Fred J. Weed, De Laval Dairy Supply Co. to W.A. Wilson, 11 December 1916; Neil S. Dow, Assistant Manager, De Laval Dairy Supply Co. to W.A. Wilson, 19 December 1916, 19 March 1917, and 16 October 1917; and Mack Robertson, President, Belleville Creamery, Ltd., Belleville, Ontario to W.A. Wilson, 1 March 1917.
17 "The Case Against Oleo" (editorial), *Farmer's Advocate*, 9 November 1916, 1835. (This is the first editorial on the subject since 2 April 1914.) "'Oleo' and 'Undesirable'" (editorial), ibid., 23 November 1916; "The Annual Cheese and Butter Makers' Meeting at Guelph," ibid., 21 December 1916, 2119; "The Oleo Interests Still Busy" (editorial), ibid., 31 May 1917, 897; and "When Oleo Comes" (editorial), ibid., 1 November 1917, 1689. "Keep Out Oleomargarine" (editorial), *London Advertiser*, 22 October 1917. Saskatchewan, Department of Agriculture, Dairy Branch, "Oleomargarine, 1916-1948" (W.A. Wilson), Dairy Commissioner, Regina to Hon. Martin Burrell, Minister of Agriculture, Ottawa, 1 December 1916; and Assistant Dairy Commissioner, Regina to J.G. Turrell, MP, 19 February 1917.
18 *Canadian Medical Association Journal* (April 1917): 324-25.
19 This pamphlet is reprinted in "Report, Returns and Statistics of the Inland Revenues ... Part III, Adulteration of Food," *Sessional Papers, 1919*, No. 14, 11-24, Bulletin No. 377, "Human Food, Considered in Its Relation to Quantity and Cost." It has not been established what circulation this bulletin had.
20 Queen's University Archives (hereafter QUA), Thomas L. Crerar correspondence, Series V, Box 178, General Subject Files, "Oleomargarine, 1917-1918."
21 *Farmer's Advocate*, 8 February 1917, 219.
22 Ibid., 29 November 1917, 1849.
23 Commons, *Debates*, 1917, 2283.
24 Saskatchewan, Department of Agriculture, Dairy Branch, "Oleomargarine, 1916-1948," J.G. Turrell to W.A. Wilson, 8 February 1917, and J.G. Turrell to F.L. Logan, 26 February 1917.
25 Ibid., D.B. Nealy, MP to W.A. Wilson, 6 February 1917.

26 Ibid., R.W. Heim, Prince Albert Creamery Co., Prince Albert, Saskatchewan to W.A. Wilson, 27 October 1917.
27 "Famine and World Hunger Are On Our Threshold," *Globe* (Toronto), 28 April 1917, 14.
28 Wilson, *History of Unilever*, 1:229.
29 "May Permit Margarine,"*Globe*, 18 October 1917, 1.
30 "Oleomargarine 1917-1918," "Memorandum re Oleomargarine," QUA, Crerar Papers, series V, Box 178, General Subject Files, undated, confidential.
31 Saskatchewan, Department of Agriculture, Dairy Branch, "Oleomargarine, 1916-1948," W.A. Wilson to J.A. McFeeter. See also "Milk Committee Favors Removal Margarine Ban," *Ottawa Farm Journal*, 16 October 1917, 6.
32 *Canada Gazette*, 25 October 1917, 2nd Extra, 1-2.
33 National Archives of Canada (hereafter NAC), W.L.M. King Papers, MG 26, J1, 76878, W.R. Motherwell to W.L.M. King, 23 May 1922.
34 "Memorandum re Proposal to Remove the Restrictions on the Sale and Manufacture of Oleomargarine in Canada," NAC, Arthur Meighen Papers, C3215, No. 2646, copy of 17 October 1917.
35 *Canada Gazette*, 17 November 1917, 2nd Extra, 1-2.
36 Ibid., 24 November 1917, 1715.
37 Ibid., 27 July 1918, 351-52.
38 "May Manufacture and Sell Oleomargarine," *Guardian* (Charlottetown), 1 November 1917, 1.
39 "Fighting the Good Shortage," *Digby Weekly Courier*, 22 June 1917, 4.
40 "Regulations Will Press Heavily on Oleomargarine" (editorial), *Maritime Farmer*, 4 December 1917, 136.
41 "May Sell 'Margarine' After November 1st," *Ottawa Evening Journal*, 25 October 1917, 3.
42 A. Fortescue Duguid, *Official History of the Canadian Forces in the Great War 1914-1919* (Ottawa: King's Printer, 1938) 1, appendix 103.
43 Interview with S.J. Harris and Major J. Mateo, Director of Food Services, both of the Department of National Defence, Ottawa, 26 November 1979.
44 Saskatchewan, Department of Agriculture, Dairy Branch, Elizabeth Shortt to E.I. Love, 8 April 1923. QUA, Crerar Correspondence, William Krogh to Crerar, 12 April 1923, 4. No reference to margarine has been found in any published memoirs such as those of Lester B. Pearson, John Diefenbaker or Arthur R.M. Lower.
45 Wilson, *History of Unilever*, 1:230, 306. Thomas Lever, one of the manufacturers of margarine in Britain, faced threats of employees wishing to leave their positions, when Lever made margarine part of his servants' meals. The employees feared poisoning or at least being degraded as human beings. Lever himself did not have it served it at his personal table.
46 J.C. Drummond and A. Wilbraham, *The Englishman's Food: A History of Five Centuries of English Diet* (London: Jonathan Cape, 1969), 438.
47 Interview with J.H. Anderson, Saskatoon, Saskatchewan, 19 July 1979.
48 "Oleomargarine Supply Is Not Very Large Yet," *Ottawa Evening Journal*, 17 December 1917, 2.
49 J. Michael Bliss, *A Canadian Millionaire: The Life and Business Times of Sir Joseph Flavelle, Bart., 1858-1959* (Toronto: Macmillan, 1978), 185.
50 Ibid., 533.
51 William Clayton, *Margarine* (London: Longmans, 1920), 35-59.
52 Johannes Hermanus van Stuijvenberg (ed.), *Margarine: An Economic, Social and Scientific History, 1869-1969* (Toronto: University of Toronto Press, 1969), 232.
53 *Sessional Papers, 1923*, No. 95, 3, and Commons, *Debates, 1923*, 3516 and 3541.
54 E.T. Love, *Oleomargarine and Its Relation to Canadian Economics* (Edmonton: Alberta Dairymen's Association, 1923), 3.
55 "P.C. 2402," *Canadian Gazette*, 30 September 1918, 1280.
56 William H. Nicholls, "Some Economic Aspects of the Margarine Industry," *Journal of Political Economy*, 54, 3 (June 1946): 228.

57 Love, *Oleomargarine and Its Relation*, 2.
58 *Globe*, 19 February 1920, 11.
59 Ibid., 24 February 1920, 11.
60 H.E. Stephenson and Carlton McNaught, *The Story of Advertising in Canada* (Toronto: Ryerson, 1940), 49-75.
61 *Globe*, 19 February 1920, 11.
62 Ibid., 26 February 1920, 11.
63 Ibid., 2 March 1920, 8.
64 Stuijvenberg, *Margarine*, 261-62.
65 The wholesale price of each food was noted in the Toronto *Globe*.
66 *Kitchener News Record*, 31 December 1917, 2.
67 Stuijvenberg, *Margarine*, 267-68.
68 *Historical Statistics* (1983), series M343-56.
69 Alberta, Department of Agriculture, *Annual Report, 1918*, 70.
70 Saskatchewan, Department of Agriculture, "Report of the Dairy Commissioner,"*Annual Report, 1917/1918*, 39.
71 "My First Bite of Oleo" (letter to the editor), *Farmer's Advocate*, 10 January 1918, 48.
72 *Maritime Merchant and Commercial Review*, 24 January 1918, 85.
73 Saskatchewan, Department of Agriculture, Dairy Branch, "Oleomargarine, 1916-1948," W.W. Moore, Vancouver to W.A. Wilson, Regina, copy, 30 October 1917, and W.A. Wilson to W.W. Moore, copy, 5 November 1917.
74 Ibid., W.A. Wilson, Regina to Neil S. Dow, Winnipeg, 23 November 1917; Dow to Wilson, 26 November 1917; and J.A. Ruddick, Ottawa to Wilson, 1 December 1917.
75 Ibid., F.H. Auld, Regina to C.F. Bailey, Assistant Deputy Minister of Agriculture, Toronto, copy, August 1918.
76 *Globe*, 8 December 1917, 11.
77 Commons, *Debates, 1918*, 288-92, 297-314.
78 Saskatchewan, Department of Agriculture, Dairy Branch, "Oleomargarine, 1916-1948," W.A. Wilson to M. Robertson, Belleville, Ontario, 8 November 1917, and Neil S. Dow to W.A. Wilson, 12 November 1917.

Chapter 4

1 *Canadian Annual Review, 1919*, 333.
2 "Annual Report of the Minister of Agriculture for the Year Ending March 31, 1921," *Sessional Papers, 1922*, No. 15, 55. See also *Historical Statistics* (1965), Series L253-60.
3 W.R. Motherwell to Mackenzie King, 22 May 1923, NAC, W.L.M. King Papers, MG 25, J1, 76872.
4 *Canadian Annual Review, 1919*, 324.
5 "The Time to Quit Dairying," *Maritime Farmer*, 25 June 1918, 583.
6 Alex McKay to Arthur Meighen, 6 February 1917, NAC, Meighen Papers, C3215, No. 2651.
7 Alberta, Legislative Assembly, *Journals, 1923*, 135-38.
8 Nova Scotia, House of Assembly, "Report of the Agricultural Committee," *Journals, 1923*, 288.
9 Five copies have been found. Nos. 1-4, in the Queen's University Archives, were published 15 December 1923, 1 and 15 January and 1 February 1924. No. 5, dated 15 February 1924 is in Saskatchewan, Department of Agriculture, Dairy Branch, "Oleomargarine, 1916-1948."
10 Published by the Alberta Dairymen's Association, March 1923.
11 Edmonton: n.p., 1924.
12 Commons, *Debates, 1923*, 3558.
13 Saskatchewan, Department of Agriculture, Dairy Branch, "Oleomargarine 1916-1948," E.T. Love to the Creamerymen of Canada, 23 May 1923.

14 Commons, *Journals, 1919*, 197.
15 Ibid., *1923*, Appendix 3, 598, "Proceedings of the Select Special Committee ... on Agricultural Conditions," testimony by E.H. Stonehouse, President, National Dairy Council of Canada.
16 Canadian Pacific, Corporate Archives, "Oleomargarine," *Agricultural and Industrial Progress in Canada*, August 1920, 136.
17 SAB, Saskatchewan, Department of Agriculture, Dairy Branch, "Oleomargarine, 1916-1948," P.F. Reed, Dairy Commissioner, Regina to F.M. Dymond, Secretary, Retail Merchants Association, Saskatoon, copy, 15 April 1920.
18 *Globe*, 19 February 1920, 11.
19 Ibid., 2 March 1920, 8.
20 *Industrial Canada*, December 1923, 62-63.
21 QUA, Crerar Papers, Series III, Box 130, "Oleomargarine, 1920-1922," Swift-Canadian Ltd. to Holmes Jowett, President, Trades and Labour Council, Lethbridge, Alberta, copy, 28 February 1920.
22 Commons, *Debates, 1923*, 3566.
23 NAC, Meighen Papers, C 3215, No. 2559-61, 14 April 1919.
24 QUA, Crerar Papers, Series III, Box 130, "Oleomargarine, 1920-1922."
25 No copies of this item have been found.
26 Ibid., issued by the Canadian manufacturers and distributors of oleomargarine, "Why Ban a Good Food?" n.d., 6.
27 Commons, "Proceedings of Select Special Committee ... on Agricultural Conditions," *Journals, 1923*, Appendix 3, 258, comment by Thomas Sales.
28 "Oleomargarine on Suspended Sentence," *Farmer's Advocate*, 13 November 1919, 2039, and Commons, *Debates, 1922*, 1810, statement by W.R. Motherwell.
29 Commons, *Journals, 1919*, Appendix 3.
30 Ibid. (second session), 77, *passim*. The act was *Statutes of Canada, 1919*, 506, 10 Geo V. cap. 24.
31 Commons, *Debates, 1919*, 7 October 1919, 890-98, 933-52, and "Country Produce, Wholesale," *Globe* (Toronto), 9 October 1919, 17.
32 Commons, *Debates, 1921*, 3871.
33 Ibid., *1922*, 1836.
34 "Dr. Tolmie Defends Action," *Daily Colonist* (Victoria), 15 February 1924, 1.
35 In 1920, 11 of 19 backbenchers were from Ontario; in 1921, 10 of 26, in 1922, 4 of 18, and in 1923, 8 of 33.
36 Commons, *Debates, 1922*, 15 May 1814 for Fielding's position and 6 June 1923, 3524 for King's. See Saskatchewan, Department of Agriculture, Dairy Branch, P.E. Reed to A.E. Potts, 24 June 1923 for a contemporary evaluation of the significance of Motherwell's work.
37 Motherwell to King, 22 May 1923, NAC, Mackenzie King Papers, MG 26, J1, No. 76872. The legislation concerning renovated butter is "An Act to Amend the Dairy Industry Act," 13-14 Geo. V, cap. 43.
38 The 1924-1925 annual report of the federal Department of Agriculture notes (p. 28) that the volume shipped to the United States was the equivalent of 24 million pounds of butter. In 1924-1925, 180,663,783 pounds of creamery butter were produced in Canada.
39 Commons, *Debates, 1923*, 3526 *passim*.
40 Province of Manitoba Archives (hereafter PAM), Papers of the United Farmers of Manitoba, Manitoba Grain Growers *Year Book 1918, 1919, 1923*.
41 PAM, Canadian Council of Agriculture Papers, Minute Books, 1911-1932 (on film).
42 John H. Thompson with Allan Seager, *Canada, 1922-39: Decades of Discord* (Toronto: McClelland and Stewart, 1985), 112.
43 Senate, *Debates, 1946*, 83.
44 "The Day at the Capital," *Winnipeg Free Press*, 16 May 1922, 1.
45 *Canadian Annual Review, 1923*, 173-78.
46 Commons, *Debates, 1922*, 1840.

47 To be classified as urban/rural the constituency had to have a minimum of 1/3 of its population in the minor classification.
48 Commons, *Debates, 1923*, 3568.
49 R. MacGregor Dawson, *William Lyon Mackenzie King, A Political Biography: 1874-1923* (Toronto: University of Toronto Press, 1958), 1:449.

Chapter 5

1 Marr and Paterson, *Canada: An Economic History*, chart 6:12; *Historical Statistics* (1965), Series A15-A19; and L.O. Stone, *Urban Development in Canada* (Ottawa: Queen's Printer, 1961), 22-34.
2 *Historical Statistics* (1965), Series L167.
3 Ibid., Series L243.
4 Ibid., Series L253-L260.
5 Ibid., Series L306-L310.
6 The price of butter declined from a pre-Depression high in 1929 of 41.9¢ wholesale to a Depression low in 1932 of 22.6¢ and then rose to a Depression high of only 28.4¢ wholesale. Ibid., Series L108; and Dennis Guest, *Emergence of Social Security in Canada*, 2nd ed., rev. (Vancouver: University of British Columbia Press, 1985), Table II, 207.
7 Interview with Dr. Douglas Busby, Orangeville, Ontario, 21 December 1980.
8 Saskatchewan, Department of Agriculture, Dairy Branch, "Oleomargarine, 1916-1948," Abbie DeLury to PC Kidd, 1 March; D'Arcy Scott to P.E. Reed, 16 March; Mrs. M.J. Robertson to Percy Reed, 15 April; P.E. Reed to Mrs. M.J. Robertson, 27 April; P.E. Reed to Abbie DeLury, 30 June; and Abbie DeLury to P.E. Reed, 2 July.
9 Commons, *Debates, 1925*, 2985. *Statutes of Canada, 1925*, 15-16 Geo. V, cap. 40, 273, "An Act to Amend the Dairy Industry Act, 1914."
10 Commons, *Debates, 1930*, 1630 and 2201. *Statutes of Canada, 1930*, 20-21, Geo. V, cap. 13, 18, "An Act to Amend the Customs Tariff."
11 Commons, *Debates, 1930* (special session), 245. *Statutes of Canada, 1930* (second session), 21 Geo. V, cap. 3, 8, "An Act to Amend the Customs Tariff."
12 John W. Warnock, *Profit Hungry: The Food Industry in Canada* (Vancouver: New Star, 1978), 93.
13 Wilson, *History of Unilever*, II, 306, 338.
14 Commons, *Journals, 1934*, 19, *passim*, and Commons, Debates, 1934, 221-24.
15 "Government Measure Roundly Scored in Legislature," *Guardian* (Charlottetown), 31 March 1937, 1. "The Dairy Industry Act," *Acts of the General Assembly of Prince Edward Island, 1937*, 2 Geo. VI, cap. 7, 84-5.
16 Saskatchewan, Department of Agriculture, Dairy Branch, "Oleomargarine, 1916-1948," Allan C. Fraser to Hon. James G. Gardiner, 29 May 1936 and P. Reed to E.T. Love, 19 September 1940.
17 Drummond, *The Englishman's Food*, 448 ff.
18 *Healthful Eating*, National Health Series No. 109 (Ottawa: King's Printer, 1944), 6.
19 *Historical Statistics* (1965) Series L167.
20 Ibid., Series L244 and L245.
21 Ibid., Series L288.
22 Royal Commission on Prices, *Report*, 1949, 47.
23 British Columbia, Legislative Assembly, *Sessional Papers, 1945*, 592, "Annual Report of Department of Agriculture."
24 *Historical Statistics* (1983), Series M363-68.
25 Commons, *Debates, 1947*, 1522.
26 PAM, Manitoba, Department of Agriculture, Milk Control Board, Correspondence, 1944-1950, Box 13, "Agricultural Food Board, Ottawa," "Report of Meeting of the Agricultural Food Board and the Provincial Milk Control Boards," 1 February 1945.
27 *Canada Year Book, 1946*, 574, and *1947*, 758.

28 "Butter Onus Laid to Board," *Globe and Mail* (Toronto), 2 December 1942, 9.
29 "Mayor Outlines Plans to Ration City's Butter," ibid., 15 December 1942, 4.
30 *Record* (Kitchener), 13 December 1942, 16.
31 *Canada Year Book, 1943-44*, 524, and *1945*, 569.
32 *Record*, 5 February 1943.
33 "Activities of Ration Board are Reviewed," *Brantford Expositor*, 3 February 1943, 7.
34 *Record*, 21 December 1942.
35 "Hint to Housewives" (editorial), *Guardian* (Charlottetown), 29 January 1948, 4.
36 "Butter Shortage Likely to Force Importation," *Winnipeg Free Press*, 12 December 1941, 8.
37 "Letters to the Editor," *Ottawa Journal*, 23 December 1942, 8. The *Journal* made no editorial comment.
38 Senate, *Debates, 1946*, 95.
39 *Expositor*, 3 February 1943.
40 "Farmers Oppose Oleomargarine: Urge Abolishing Daylight Saving," *Vancouver News Herald*, 24 November 1944, 5.
41 Saskatchewan Department of Agriculture, Dairy Branch, "Oleomargarine 1916-1948," P. Reed to E.T. Love, 19 September 1940.
42 "Dieppe," CBC-TV program presented on 12 November 1979.
43 W.R. Feasby (ed.), *Official History of the Canadian Medical Services: 1939-1945*, 2 vols. (Ottawa: Queen's Printer, 1953), 2:156.
44 Farley Mowat, *And No Birds Sang* (Toronto: McClelland and Stewart, 1979), 164.
45 Bing Coughlin, *Herbie* (Toronto: Thomas Nelson, n.d.), 122.
46 Interview with William Hayward, Waterloo, Ontario, 16 December 1980; Canadian Legion members to W.H. Heick correspondence, Mrs. Sheila Davidson, Seton Portage, B.C. to author, 15 February 1982.
47 Interviews with Carl Schlote, Kitchener, Ontario, Arthur Schlote, Kitchener, Ontario, and William Schlote, Waterloo, Ontario, 22 June 1981.
48 Ross Munro to author, 7 August 1981.
49 Ken Slade, Mascouche Heights, Quebec, to author, 15 June 1983.
50 F.F. Tisdale, "Further Report on the Canadian Red Cross Food Parcels for British Prisoners-of-War," *Canadian Medical Association Journal*, 55, 3 (February 1944): 135-38. F.F. Tisdale et al., "Final Report on the Canadian Red Cross Food Parcels for Prisoners-of-War," ibid., 59, 4 (March 1949): 270-86.
51 D.G. Dancocks, *In Enemy Hands: Canadian Prisoners of War, 1939-45* (Edmonton: Hurtig, 1983), 83.
52 John Mellor, *Forgotten Heroes, The Canadians at Dieppe* (Toronto: Methuen, 1975).
53 Saskatchewan, Department of Agriculture, Dairy Branch "Oleomargarine 1916-1948," P. Reed to S.C. Burton, 16 October 1940.
54 Mrs. Sheila Davidson, Seton Portage, B.C., to author, 19 January 1982.
55 Mrs. Margaret Doucette, Whitby, Ontario, to author, 16 January 1982.
56 Mrs. J.M. Zarn, Sicamous, B.C., to author, 30 November 1981.
57 Mrs. Olga Rains, North Bay, Ontario, to author, 21 December 1981.

Chapter 6

1 Mackenzie King Papers, MG26, Vol. 398, 360107, Minute of a meeting of a delegation representing the Canadian Federation of Agriculture with Ministers of the Government, 28 March 1946.
2 R.D. Francis and D.B. Smith, *Readings in Canadian History: Post-Confederation* (Toronto: Holt, Rinehart and Winston, 1986), 502.
3 *Historical Statistics of Canada* (Ottawa: Statistics Canada, 1983), series A250, A252, A256, A258, B76, B81.
4 Ibid. (Toronto: Macmillan, 1965), series D1.

5 *Farmer's Advocate*, 22 January 1948, 55.
6 Ibid., 28 February 1946, 119.
7 "World Dairy Production," *Halifax Herald*, 6 September 1948, 4.
8 "What Will People Eat Tomorrow?" *Farmer's Advocate*, 10 April 1947, 262.
9 Saskatchewan, Department of Agriculture, "Report of Dairy Commissioner," *Annual Report, 1946* 67.
10 "Creamery Butter Off 20 Percent," *Globe and Mail*, 11 April 1946, 2.
11 Senate, *Debates, 1946*, 13 April 1946, 88.
12 "Butter," *Canadian Forum*, February 1947, 244.
13 "Low Prices of Milk Forces Herd Exports," *Globe and Mail*, 2 May 1946, 7.
14 "This is Absolute Monopoly," *Halifax Herald*, 19 January 1948, 4, twelve cents per pound compared to production costs of 23¢ per pound.
15 NAC, William Daum Euler Papers, Euler to Charles Black, 10 April 1946.
16 Lawrence Goodwyn, *The Populist Moment* (New York: Oxford, 1978), xix, and Dave McIntosh, *The Collectors: A History of Canadian Customs and Excise* (Toronto: NC Press, 1984), 145.
17 Euler Papers, Euler to Altman, 9 December 1947.
18 Ibid., Euler to James Eckman, Vancouver, 16 May 1946.
19 Roslyn Cluett, "The Role of Newspapers in the National Margarine Debate, 1946-50," unpublished paper written for History 680 at Wilfrid Laurier University, 1981-82, 36-37.
20 "Senate Rejects Margarine Bill," *Kitchener-Waterloo Record*, 9 May 1946, 22.
21 Euler Papers; Legion/Heick correspondence; and "Oleo Sale Urged by Vet's Parley," *Vancouver News-Herald*, 30 October 1947, 2.
22 In Euler Papers.
23 Saskatchewan, Department of Agriculture Dairy Branch. "Oleomargarine 1916-1948", H.S. Hanna to D.M. Allan, 27 May 1946.
24 Editorial, *Farmer's Advocate*, 9 May 1946, 351.
25 "Watch Out for New Zealand Butter," ibid., 23 January 1947, 43.
26 Senate, *Debates, 1947*, 172 and 174.
27 "Senate Again Beats Oleomargarine Bill," *Record*, 28 March 1947, 13.
28 "Will Try to Prove Ban on Margarine is *Ultra Vires*," *Globe and Mail*, 25 June 1947, 2. No evidence as to how far this action was taken has been discovered.
29 "City-Versus-Country Split Develops in Liberal Ranks on Oleomargarine," *Daily Colonist* (Victoria), 6 February 1948, 1.
30 For example, *Ottawa Daily Citizen*, 26 March 1947.
31 Senate, *Debates, 1947*, 167 and Euler Papers.
32 Helen Jones Dawson, "The Consumers Association of Canada," *Canadian Public Administration*, 6, 1 (1963): 92-118.
33 "The Margarine Trend," *Calgary Daily Herald*, 6 October 1948, 3.

The April 1948 Gallup Poll Results

Question: At the present time it is against the law to sell oleomargarine. Would you like to see margarine sold by the stores, or do you approve of the present ban?

	All Respondents			Farm Votes Only			Non-Farm Votes Only		
	Feb '43	Apr '47	Today	Feb '43	Apr '47	Today	Feb '43	Apr '47	Today
Favour Sale	35%	45%	58%	25%	27%	34%	38%	49%	67%
Oppose Sale	45	40	29	54	55	54	42	36	20
Undecided	20	15	13	21	18	12	20	15	13

54% of Canadians never used margarine.

	Have Used It	Haven't Used It
Favour Sale	71%	49%
Oppose Sale	24%	33%
Undecided	5%	18%

34 "Margarine," *Canadian Medical Association Journal*, 57, 10 (August 1947): 169.
35 "Seeks Provision of Margarine," *Daily Colonist* (Victoria), 29 August 1947, 3.
36 "Discontinue Fight for Margarine Production," *Manitoba Cooperator*, 25 December 1947, 6.
37 Euler Papers, Euler to Senator James Murdock, 13 November 1947.
38 Senate, *Debates, 1947*, 234.
39 "No Change Noted in Butter and Eggs," *Globe and Mail*, 3 July 1947, 19, and "Butter Supplies in Good Demand," ibid., 29 November 1947, 21.
40 *Canadian Dairy and Ice Cream Journal* (December 1947): 58.
41 NAC, Canadian Federation of Agriculture Papers, vol. 57, File 87 a,b,d,g,h,i and File 88.
42 Mackenzie King Papers, vol. 424, 384750, 19 February 1947, G.J. Matte, Private Secretary to Hon. J.G. Gardiner, Minister of Agriculture.
43 "Butter vs. Margarine," *Globe and Mail*, 28 November 1947, 11.
44 "So Let Them Eat Cake" (editorial), ibid., 2 December 1947, 6.
45 Editorial, *Canadian Dairy and Ice Cream Journal* (September 1947): 56.
46 Robert Bothwell, Ian Drummond, and John English, *Canada Since 1945: Power, Politics and Provincialism*, (Toronto: University of Toronto Press, 1981), 90.
47 *Historical Statistics* (1983), series K8.
48 Ibid., series L243, L167, L244.
49 Ibid., series M36, M368, M434, and M438.
50 Ibid., series M434 and M438.
51 Commons, *Journals, 1948*, 725.
52 *Historical Statistics* (1965), series I 108.
53 "Prices Push Incomes Below Family Health Danger Line," *Globe and Mail*, 8 January 1948, 1.
54 "Food Alone Now Costs More than Food, Fuel, Rent," ibid., 9.
55 "Argues Margarine Robs Dairymen," ibid., 17 December 1947, 2.
56 "Find Families' Income $1820 Can't Buy as Much Butter," *Toronto Star*, 21 February 1948, n.p.
57 "Week to See Testing Consumer Resistance to Soaring Food Prices," *Globe and Mail*, 12 January 1948, 2.
58 "Here's How to Make Butter Go Further," ibid., 12 April 1948, 12.
59 Ibid., 22 April 1948, 2.
60 "Get Pound of Butter with $7.00 Purchase," *Halifax Herald*, 1 April 1948, 1.
61 "Ottawa Glad Food Prices Drain Cash," *Globe and Mail*, 10 January 1948, 15.
62 "70 Cent Butter Monday," *Leader Post* (Regina), 18 January 1948, 1.
63 "That Nasty Word Again," ibid., 15 April 1948, 15.
64 "Professor Names Head of Commission on Prices," *Herald*, 9 July 1948, 1.
65 "Butter Famine in Winter Forecast if Ottawa Fails to Take Action Speedily," *Globe and Mail*, 8 July 1948, 3.
66 "Butter Increase Cent: Maritime Price Highest," *Herald*, 21 July 1948, 1.
67 "Blames Butter Shortage on Cost of Farm Labour," ibid., 31 July 1948, 1.
68 "Find Families' Income $1820 Can't Buy as Much Butter," *Star*, 21 February 1948, n.p.; "They Call It Political Dynamite," *Herald*, 14 February 1948, 4; and "Urges Use of Margarine," ibid., 16 December 1948, 4.
69 Interview with Dr. Hans Heick, Ottawa, 16 August 1981.
70 "Urge All Urban Municipalities to Demand Oleo," *Star*, 13 April 1948, n.p.

71 Editorial, *Farmer's Advocate*, 12 February 1948, 73.
72 "Margarine Resolution is Vetoed," *Herald*, 6 August 1948, 5.
73 "Margarine Favoured by Party," ibid., 4 October 1948, 1.
74 "Says Margarine," *Guardian* (Charlottetown), 9 September 1948, 1.
75 "Margarine Favoured by Party," *Herald*, 4 October 1948, 1.
76 Owen Carrigan, compiler, *Canadian Party Platforms 1867-1968* (Toronto: Copp Clark, 1968), 177.
77 Canadian Federation of Agriculture Papers, series 3, vol. 56, Files 87 c,d,e,f,i and series 88, vol. 57, Files a and b.
78 *Maritime Cooperator*, 15 February 1948, 4.
79 NAC, Ian Mackenzie Papers, vol. 46, file C.N.S. 7, J.S.E. McCague, President, Dairy Farmers of Canada to Senator Ian McKenzie, 12 April 1948. Copy is in Euler Papers.
80 "Drive Aimed at Margarine," *Leader-Post*, 17 January 1948, 5.
81 Saskatchewan, Dept. of Agriculture, Dairy Branch, "Oleomargarine, 1916-1948," Earle Kitchen to W.[sic]S. Hanna, 30 January 1948.
82 Ibid., C.H.P. Killick, to H.S. Hanna, 11 October 1947.
83 Senate, *Journals, 1948*, 12, and Commons, *Journals, 1948*, 56.
84 Commons, *Debates, 1948*, 370. Sinclair to author, 5 January 1977.
85 "Love, Sinclair Debate Margarine Question," *Cooperator*, 4 February 1948, 6.
86 "New York State Legislature Adopts Bill Suspending Ban for Year," *New York Times*, 10 February 1948, 4, 5, and "Oleo Compromise Bill Signed in New York State," *Montreal Gazette*, 19 February 1948, 1.
87 "Allowed in New Jersey," *Winnipeg Free Press*, 23 April 1948, 4.
88 Senate, *Debates, 1948*, 176 passim.
89 "Ignoring Public Opinion" (editorial), *Leader-Post*, 12 May 1948, 11.
90 Senate, *Debates, 1948*, 493.
91 Senate, *Journals, 1948*, 382.
92 Senate, *Debates, 1948*, 554.
93 Mackenzie King Diary, 19 March 1948, T.S. 240/229-230, 1.
94 "Sinclair Promises Action to Allow Margarine Sale," *Vancouver Sun*, 18 September 1947, 1, and "Butter," *Canadian Forum*, February 1947, 244.
95 *Time Magazine*, 5 January 1948, 22.
96 Mackenzie King Papers, MG 26, J4, V. 310, File 3294, Oleomargarine, 1948, P.C. 214236, "Memorandum for the Prime Minister: re Statement on Oleomargarine at Caucus," confidential, 3 February 1948.
97 "Hon. Gregg on Oleomargarine," *Lethbridge Herald*, 26 April 1948, 2,
98 "Margarine Use Bad for Health, Gardiner Holds," *Globe and Mail*, 6 April 1948, 1; also "Buttering the Ballots" (editorial), ibid., 6 April 1948, 6.
99 Commons, *Debates, 1948*, 4276.
100 St. John Chadwick, *Newfoundland: Island Into Province* (Cambridge: Cambridge University Press, 1967), 257.
101 "Margarine for Canadians" (editorial), *Evening Telegram* (St. John's), 16 December 1948, 6.
102 "Wasted Advice – and Money," *Halifax Herald*, 13 January 1948, 4.
103 Commons, *Debates, 1948*, 2719-25, and "Butter Premium Plan Probed by Committee," *Halifax Herald*, 7 April 1948, 3.
104 Commons, *Debates, 1948*, 3152 passim.
105 Ibid., and Commons, *Journals, 1948*, 500.
106 Commons, *Debates, 1948*, 3185.
107 "Laid 'Oscar' for Jim – But No Marge – Valley MP," *Vancouver Daily Province*, 21 April 1948, 1.
108 "Cheese Board Ready to Accept Margarine, Ask Export Concession," *Globe and Mail*, 8 July 1948, 15.
109 "A Floor Price for Butter," *Farmer's Advocate*, 10 February 1948, 81.
110 "Dairy Council Admits Need for Margarine," *Globe and Mail*, 11 August 1948, 1.
111 "This in Contrast to 25% in the U.S.A.," *Star*, 16 September 1948, n.p.

112 Copy in Euler Papers.
113 "And the 'Famine' Continues," *Globe and Mail*, 30 September 1948, 6.
114 "68% Favour Margarine," *Kitchener-Waterloo Record*, 6 October 1948, 6.

Chapter 7

1 The essence of this chapter is based upon the excellent cognate essay written by John Gaskell in 1982-1983 as a requirement for the master's degree at Wilfrid Laurier University. It is entitled "The Supreme Court of Canada and section 5 (a) of the Dairy Industry Act."
2 "Reference re validity of section 5 (a) of the Dairy Industry Act," in *Canada Law Reports*, Part 1, 1949, 2.
3 "Margarine Use Crime, Deputy Minister Tells Supreme Court Judges," *Globe and Mail*, 6 October 1948, 3.
4 "Milliken Backs Oleo Argument," *Leader-Post* (Regina), 6 October 1948, 1.
5 "The Legal Lady," *Maclean's*, 15 July 1949, 15.
6 [1949] *Supreme Court Reports*, 1, 1-89.
7 Ibid., 21.
8 Editorial, *Calgary Herald*, 15 December 1948, 4.
9 "Citizens Hail Margarine Decision, Removal of Ban Approved," *Winnipeg Free Press*, 15 December 1949, 1.
10 "Lifting of Margarine Ban Called Betrayal," *Brantford Expositor*, 16 December 1948, 16.
11 "Swan Song for Ban," ibid., 17 December 1948, 4.
12 "Butter Producers Irked as Margarine Ban Lifted," *Globe and Mail*, 15 December 1948, 15.
13 "Oleo Injunction Report Denied," *Kitchener-Waterloo Record*, 17 December 1948, 7.
14 "Cabinet Ponders Margarine Issue," *Leader-Post*, 3 January 1949, 2.
15 "Appeal on Oleo Ruling Unlikely: St. Laurent," *Globe and Mail*, 24 December 1948, 3.
16 Papers of the Dairy Farmers of Canada, R. Stanley to Earle Kitchen, 31 December 1948.
17 Ibid., H.H. Hannam to F.P. Varcoe (copy), 8 August 1949.
18 Euler Papers, Euler to Hayden, 5 July 1949.
19 Papers of the Dairy Farmers of Canada, R.H. Milliken to Earle Kitchen, 22 December 1950, and Kitchen to Milliken, 4 September 1951. Respective shares for each party are not known.
20 "Decision Reserved by Privy Council on Oleo Appeal," *Globe and Mail*, 30 June 1950, 10.
21 "Reference re Validity of Section 5 (a) of the Dairy Industry Act, R.S.C. 1927, Chapter 45," *Canada Law Reports*, Part I, 1949, 40-89.
22 "Vancouver Plant Turns Out First Batch of Oleo," *Record*, 18 December 1948, 12.
23 In 1989 the margarine industry determined to use the courts one more time. The Institute of Edible Oil Foods argued that the Charter of Rights gave the Canadian consumer the right to eat any healthful food. In September 1990 the Court declined to hear the case.

Chapter 8

1 *Census, 1951*, IV, Table 16 and VI, Table 64; *Historical Statistics* (1965), Series L73-77.
2 *Census, 1951*, VI, Table 64.
3 *Farmer's Advocate*, 22 January 1948, 34.
4 "A Cow Per Family" (editorial), *Guardian* (Charlottetown), 9 October 1948, 4.
5 Euler Papers, C.D. Howe to Euler, 15 December 1949.
6 "U.S. Farm Groups Give In," *Globe and Mail*, 22 February 1949, 6.

7 "Citizens Forum: High Cost of Living," *Food for Thought*, February 1951, 32-33.
8 Senate, *Debates 1951*, 722.
9 Privy Council 1953 – 241/857.
10 Don Mitchell, *Politics of Food* (Toronto: James Lorimer, 1975), 121.
11 "Minister of Margarine" (editorial), *Globe and Mail*, August 9, 1951, 6.
12 Commons, *Debates 1949*, 17 February 1949, 617.
13 Ibid., *1951*, 25-26 June 1951, 4067 *passim*.
14 Senate, *Debates 1951*, 28 June 1951, 719 ff. *Statutes of Canada 1950-51*, 15 Geo. VI, cap. 39, 225-30.
15 "Unfair and Dangerous Law" (editorial), *Toronto Star*, 25 August 1951, n.p.
16 "Is Margarine Outside the Constitution?" (editorial), *Maclean's*, 15 August 1951, 2.
17 "Ottawa to Give Up Plans to Control Margarine," *Financial Post*, 21 June 1952, 24.
18 Commons, *Debates 1952*, 3245 *passim*; Senate, *Debates 1952*, 9 *passim*; and "An Act to Amend the Canada Dairy Products Act," *Statutes of Canada, 1952*, 61, 1 Eliz. II, cap. 16.
19 J.H. Duplan, President, National Dairy Council to Louis St. Laurent, 24 January 1949, SAB, T.C. Douglas Papers.
20 Douglas Campbell to T.C. Douglas, 16 February 1949, ibid.
21 Louis St. Laurent to T.C. Douglas, 11 March 1949, ibid.
22 Alberta, Department of Agriculture, "Meeting of Federation of Dairy Farmers and Provincial Departments of Agriculture," Ottawa, 14 March 1949, 72.302 Ministers' and Deputy Ministers' Files C 1937-1970, File 279, Oleomargarine, 1949.
23 Saskatchewan, Legislative Assembly, *Debates and Proceedings, 1949*, 2 April 1949, 939-41.
24 *Statutes of British Columbia*, 25 Geo. V, cap. 16, 909.
25 "Province May Prohibit Manufacture and Sale of Margarine in P.E.I." *Guardian* (Charlottetown), 31 December 1948, 1.
26 Ibid., 31 December 1948, 2.
27 "Butter vs. Margarine," ibid., 17 January 1948, 4.
28 "Margarine Ban Lifted" (editorial), ibid., 16 December 1948, 4.
29 "P.E.I. Dairymen Optimistic: Make Plans for Expansion," ibid., 24 February 1948, 1.
30 "Ban Proclaimed on Margarine by Lt. Governor," ibid., 14 January 1949, 1. *Acts of the General Assembly of Prince Edward Island, 1937*, 1 Geo. VI, cap. 7, 84-85.
31 P.E.I., *Journal of the Legislative Assembly of the Province of Prince Edward Island, 1950*, 108-54, and *Acts of the General Assembly of Prince Edward Island, 1950*, 14 Geo. VI, cap. 12, 67-70.
32 Interview with Mrs. Fran Bowden, Charlottetown, 8 October 1980. (No evidence of any charges having been laid has been uncovered.)
33 "Manufacture of Oleo Must Await Quebec Decision, Duplessis Warns," *Gazette* (Montreal), 18 December 1948, 1.
34 Euler Papers, C. Stanley to W.D. Euler, 1 February 1949.
35 "Margarine, 1948-53," NAC papers of the Montreal Council of Women, V II, dated 13 January 1949, 23 February 1949 and 5 March 1949, File 25, MG 28-1-164.
36 Canadian Institute of Public Opinion, "Margarine Sale in Quebec Urged by 55% of Province's Voters," *The Gallup Report*, 28 April 1949.
37 "Butter Price Affected by Margarine," *Montreal Daily Star*, 3 March 1949, 1.
38 *Statutes of the Province of Quebec, 1949*, 13 Geo. VI, cap. 44, 167-69.
39 "Barre Claims Oleo Would 'Disgust' Buyer," *Montreal Daily Star*, 4 March 1949, 1.
40 "'Marge Squad' Set to Swing into Action," ibid., 7 March 1949, 1.
41 "Quebec Law Bans Sale of Margarine," ibid., 19 March 1949, 1.
42 "Queen's Park Views Oleo with Hands-Off Feeling," *Globe and Mail*, 17 December 1948, 3.
43 Euler Papers, C. Stanley to W.D. Euler, 1 February 1949.
44 Public Archives of Ontario (hereafter PAO), Ontario Department of Agriculture Papers, Rb 4, series A-2, 40-1, 1948-49, 15 December 1949, C.A. Massey to Leslie Blackwell.
45 New Brunswick, *Report of the Department of Agriculture for the Province of New Brunswick, 1949*, 53-4.

46 *Statutes of the Province of Ontario, 1949*, 13 Geo. VI, cap. 66, 331-32. PAO, Ontario Department of Agriculture, RG 3, E, 15-A, brief to the Premier and members of the Cabinet, from the Ontario Federation of Agriculture.
47 "79-2 for Pale Oleo," *Globe and Mail*, 29 March 1949, 2.
48 "The Edible Oil Products Acts," *Statutes of the Province of Ontario, 1952*, 15 Geo. VI, cap. 26, 111-12.
49 Ibid., *1953*, 1 Eliz. II, cap. 31, 259-60.
50 "Dairymen Discuss Margarine Issue," *Telegraph-Journal* (Saint John), 2 April 1949, 3; New Brunswick, *Report of the Department of Agriculture for the Province of New Brunswick, 1949*, 53-54; and "Urge Margarine Sale be Allowed," *Halifax Chronicle-Herald*, 11 April 1949, 1.
51 *Report of the Department of Agriculture for the Province of New Brunswick, 1949*, 53-54.
52 *Journals of the Legislative Assembly of New Brunswick, 1949*, 165 *passim*, and *Acts of the Legislature of New Brunswick, 1949*, 13 Geo. VI, cap. 23, 187-89. In 1951, as part of a general revision of the statutes, the legislature passed cap. 191. No fundamental changes of cap. 23 were made (*Synoptic Report of the Proceedings of the Legislative Assembly of New Brunswick, 1951*, 215; *Acts of the Legislature of New Brunswick, 1951*, 15 Geo. VI, cap. 191, 1019-21).
53 *Journals of the Legislative Assembly of the Province of New Brunswick, 1950*, 147 *passim*, and *Acts of the Legislature of New Brunswick, 1950*, 14 Geo. VI, cap. 51, 115.
54 "The Record is Plain" (editorial), *Chronicle-Herald*, 1 May 1948, 4.
55 "A Gratuitous Attack" (editorial), ibid., 21 February 1948, 4.
56 "How Not to Do It" (editorial), ibid., 17 December 1948, 4.
57 Nova Scotia, *Journals and Proceedings of the House of Assembly, 1949*, 87 *passim*, and *Statutes of the Province of Nova Scotia, 1949*, 13 Geo. VI, cap. 3, 38-41.
58 "Clash Margarine Question," *Chronicle-Herald*, 12 April 1949, 1.
59 "Quite an 'Ad' for Margarine" (editorial), ibid., 13 April 1949, 4.
60 "Whither Dairying?" (editorial), *Manitoba Cooperator*, 23 December 1948, 4.
61 "Oleomargarine" (editorial), ibid., 10 February 1949, 4.
62 "Manitoba's Women's Institute Convention Backs Margarine," ibid., 24 June 1948, 9.
63 Manitoba, Legislative Assembly, *Journals, 1949*, 15 February 1949, 28 *passim*.
64 Manitoba, *Journals, 1949*, and *Statutes of Manitoba, 1949*, 13 Geo. VI, cap. 35, 107-09.
65 "Margarine Discussed," *Leader-Post* (Regina), 18 January 1949, 3, and Saskatchewan, *Journals of the Legislative Assembly, 1949*, 60 *passim*.
66 Saskatchewan, Legislative Assembly, *Debates, 1949*, 3 March 1949, 837 *passim*.
67 T.C. Douglas Papers, Brief of Saskatchewan Dairy Association to Hon. I.C. Nollet, Minister of Agriculture, n.d.
68 Saskatchewan, *Debates, 1949*, 2 April 1949, 939 *passim*, and *Statutes of the Province of Saskatchewan, 1949*, 13 Geo. VI, cap. 77, 515-17.
69 T.C. Douglas Papers (copy), Douglas to Rev. J. Mayne, 23 May 1949, and (copy) I.C. Nollet to Rev. J. Mayne, 31 May 1949.
70 "Gas Line Bribe Charge Ruled Out in House," PAA, Alberta Scrapbook Hansard, Reel 14, exposure 110, 2 March 1950.
71 D.H. McCallum to C.E. [sic] Manning, "Oleomargarine, 1947," PAA, Alberta, Department of Agriculture, Ministers' and Deputy Ministers' Files, c. 1939-70, File 19.
72 "Dairymen Ask Ceiling Be Lifted on Butter," *The Albertan* (Calgary), 11 February 1949, 1.
73 S. F. McDougall to David A. Ure, "Oleomargarine, 1949," Alberta, Department of Agriculture, Ministers' and Deputy Ministers' Files, c. 1949-70, File 270.
74 I.A. McArthur to E. C. Manning, "Agricultural Products, 1949-1955," PAA, 62.389, Premiers' Papers, File 1529.
75 L.W. Munroe to Hon. D. Ure, "Oleomargarine, 1949," Department of Agriculture, Ministers' and Deputy Ministers' Files, c. 1939-1970, File 279.
76 PAA, Alberta Federation of Agriculture, Item 91a, Dairying 1949-70, 7-4-1, Dairying 1949-53.
77 "Yellow Margarine Ban Sought by Farm Group," *Calgary Herald*, 21 March 1949, 1.
78 "Butter-Colored Oleo is Banned in Alberta," ibid., 26 March 1949, 1.

79 "House Prorogued, Bans Yellow Oleo," *Alberta Scrapbook Hansard*, Reel 15, exposure 136, 27 March 1949, and *Statutes of the Province of Alberta, 1949*, 13 Geo. VI, cap. 62, 287-88.
80 Alberta, Legislative Assembly, *Journals, 1950*, 20 *passim*; "Gas Line Bribe Charge Ruled Out in House," *Alberta Scrapbook Hansard*, Reel 15, exposure 110, 2 March 1950; and *Statutes of the Province of Alberta, 1950*, 14 Geo. VI, cap. 39, 235.
81 "Dairymen's Margarine Stand Rapped," *Vancouver Sun*, 3 January 1949, 3.
82 British Columbia, Legislative Assembly, "Annual Report of the Minister of Agriculture," *Sessional Papers, 1949*, R. 14.
83 "Margarine Needlessly Delayed" (editorial), *Vancouver Sun*, 29 December 1948, 4.
84 "Further Oleo Curbs Demanded in House," *Vancouver News-Herald*, 2 March 1949, 5.
85 "Oleo Lobby Meets Stiff Resistance," ibid., 31 January 1949, 2.
86 "Penny Starts Another Fight for Margarine," *Vancouver Sun*, 29 December 1949, 18, and "Mr. Wismer Don't Be Mean – Lift That Ban on Margarine," *Daily Colonist* (Victoria), 9 January 1949, 3.
87 "Urge Manufacture and Sale of Margarine in Province," *Victoria Daily Times*, 11 January 1949, 6.
88 "Province Has Power to Free Margarine Now," ibid., 30 December 1948, 4.
89 "Poll Shows Margarine Not Opposed," ibid., 15 January 1949, 2.
90 "Woman MLA Demands Color for Margarine," *Vancouver News-Herald*, 16 February 1949, 2.
91 British Columbia, Legislative Assembly, *Journals, 1949*, 105 *passim*; *Statutes of British Columbia, 1949*, 13 Geo. VI, cap. 16, 41; and ibid., 13 Geo. VI, cap. 48, 141-42.
92 "'White' Oleo to Spark Colorful House Fight," *Vancouver News-Herald*, 18 November 1949, 3.
93 "Margarine Color Ban Upheld in Legislature," ibid., 13 April 1951, 18, and British Columbia, Legislative Assembly, *Journals, 1951*, 79 and 95 *passim*.
94 "NWT Margarine Rules to Follow Any and All," *Edmonton Bulletin*, 28 October 1949, 16.

Chapter 9

1 Canada, Department of Health, *Food and Drug Regulations*, B.09.016-B.09.022.
2 "Breslau Couple Caught by 'Forgotten' Margarine Import Ban Enacted in 1903," *Kitchener-Waterloo Record*, 17 May 1984, B11.
3 M.A. Cohen, Deputy Minister of Finance to author, 17 August 1984.
4 "CAC Asks Ottawa to Cut Tax on Margarine," *Globe and Mail*, 14 June 1963, 10.
5 "Newfoundland Escaped It, Now All Canada May See 'Marge' Tax Dropped," *Financial Post*, 17 April 1971, 40.
6 Senate, *Debates, 1960*, 860-61 and 962.
7 "Margarine Men Muster Forces Against Tax," *Financial Post*, 12 May 1962, 11.
8 Bruce West, "Margarine Lament," *Globe and Mail*, 25 May 1971, 31.
9 Institute of Edible Oil Foods, "Proposal to the Hon. E.J. Benson, PC, MP Minister of Finance for the Removal of the 12% Sales Tax on Margarine" (February 1971).
10 Canada, Royal Commission on Taxation, *Summary of Public Hearings, 1963-64*, 39, and ibid., *Report of Royal Commission on Taxation*, V, *Sales Tax and General Tax Administration* (1966), 175.
11 Canada, Parliament, *Proceedings of the Special Joint Committee of the Senate and House of Commons on Consumer Credit, 1966-67*, No. 41 *Progress Report* (1967), 3458-59.
12 "Newfoundland Escaped It, Now All Canada May See 'Marge' Tax Dropped," *Financial Post*, 17 April 1971, 40.
13 "Dairies Ripped Off by Using Lard in Butter," *Winnipeg Free Press*, 17 March 1977, 26.
14 Canada, House of Commons, *Sessional Papers, 1953-54*, no. 204A, R. 14, D2, vol. 613, not printed.
15 "$500,000 Fraud Hits 'Butter' Buyers," *Globe and Mail*, 13 January 1983, 4.

16 Newfoundland, Food and Drug (Margarine) *Regulations, 1966.*
17 "PEI Dairymen Deny Butter Is Too Costly in Margarine Tiff," *Financial Post*, 20 March 1965, 37.
18 "Why Margarine Makers Hit by 10 Different Headaches," ibid., 20 March 1965, 5.
19 Prince Edward Island, Legislative Assembly, *Journals, 1965, 36 passim.*
20 *Acts of the General Assembly of Prince Edward Island, 1965*, 14 Eliz. II, cap. 16, 68-70.
21 Ibid., 23 Eliz. II, cap. 78, 303. Prince Edward Island, Legislative Assembly, *Journals, 1974, 168 passim.*
22 "Duplessis Admits Oleo Ban Less Effective than Desirable," *Montreal Star*, 12 January 1952, 7.
23 "Duplessis Margarine Spy Law Protested by CCF in Quebec," *Globe and Mail*, 10 November 1952, 14.
24 "Upper House Amendments Tighten Oleo Law," *Montreal Star*, 18 December 1953, 49, and *Statutes of the Province of Quebec, 1953/54*, 2-3 Eliz. II, cap 6, 31-34.
25 "Quebec Comes of Age" (editorial), *Globe and Mail*, 25 January 1961, 6.
26 Quebec, Legislative Assembly, *Journals, 1969*, 814-1003 *passim*, and *Statutes, 1969*, 9-10, Eliz. II, cap. 59, 309-12.
27 Quebec, *Journals, 1969*, 653-731, and *Statutes, 1969*, 18 Eliz. II, cap. 45, 273-92.
28 Institute of Edible Oil Foods, Files, Proposal to the Hon. Normand Toupin, Minister of Agriculture for Quebec by Standard Brands Ltd. and Kraft Foods Ltd., 19 November 1971, 4.
29 Ibid., copy, Leonard Roy, Executive Vice-President, *Le Conseil de l'Industrie Laitière du Québec*, Montreal to Honourable Normand Toupin, Minister of Agriculture, Quebec, 5 August 1970.
30 Ibid., Leonard Roy to Honourable Robert Bourassa, 17 August 1978 and copy, G.G.E. Steele to R.S. Hurlbut, 18 August 1970.
31 Ibid., Standard Brands/Kraft proposal, 19 November 1971.
32 "Dairy Products and Dairy Products Substitutes, Regulation Respecting Substitutes (Amendment)," *Gazette Officielle du Québec*, No. 51, 23 December 1972, 11530, O.C. 3768-72, 13 December 1972.
33 Nova Scotia, *Journals and Proceedings of the House of Assembly, 1961*, 19, 32, and "Would Allow Coloring in Margarine," *Chronicle-Herald*, 9 February 1961, 1.
34 *Journals and Proceedings of the House of Assembly, 1962*, 14 *passim*, and "Margarine Bill Goes Through—Without Comment," *Chronicle-Herald*, 4 April 1962, 3.
35 Nova Scotia, *Debates of the Nova Scotia House of Assembly, 1972*, 1743 *passim*, and *Statutes of Nova Scotia, 1972*, 21 Eliz. II, cap. 42, 407.
36 Saskatchewan, Legislative Assembly, *Debates 1974*, 865.
37 Nova Scotia, *Debates, 1974*, 690 *passim*, and *Statutes, 1974*, 23 Eliz. II, cap. 19, 124.
38 H.D. Hopkins, Common Clerk, Saint John, to C.B. Sherwood, Minister of Agriculture, "Dairying," Public Archives of New Brunswick (hereafter PANB), Department of Agriculture, R.S. 124, RAG, 1952-1953, 11 February 1953; ibid., J.K. King to J.R. Bell, 23 February 1953; Lillian Queen to Donald Patterson, 13 April 1953; ibid., Mildred Underhill to C.B. Sherwood, 30 March 1953; and ibid., Jessie Lawson to C.B. Sherwood, 7 February 1953.
39 Department of Agriculture, "Dairying," brief by Southern N.B. District Farmers' Federation to Cabinet of New Brunswick Government, April 1953, PANB, R.S. 124, RAG, 1952-54; see also ibid., C.B. Sherwood to W. Wilson Weldone, 5 August 1953, and ibid., bulletin of the N.B. Dairymen's Association, 20 October 1953.
40 New Brunswick, Legislative Assembly, *Journal, 1954*, 169-98 *passim*, and *Acts, 1954*, 3 Eliz. II, cap. 65, 155.
41 New Brunswick, *Journals, 1965*, 51 *passim*, and *Acts, 1965*, 14 Eliz. II, cap. 43, 78.
42 New Brunswick, *Journals, 1968*, 22 *passim*, and *Acts, 1968*, 17 Eliz. II, cap. 43, 345.
43 New Brunswick, *Journals, 1973*, 90 *passim*; *Acts, 1973*, 22 Eliz. II, cap 66, 281.
44 New Brunswick, *Journals, 1977*, 37 *passim*; *Synoptic Report of the Proceedings of the Legislative Assembly of New Brunswick, 1977*, 165, and *Acts, 1977*, 26 Eliz. II, cap. 37, n.p.
45 "Manitoba May O.K. Yellow Margarine," *Financial Post*, 30 January 1960, 17.

46 "Rural Manitoba Favours Margarine Over Butter," *Winnipeg Tribune*, 4 October 1952, 1, and "Rural Manitoba and Colored Margaine" (editorial), ibid., 6 October 1952, 6.
47 PAM, Manitoba Enquiry Commission, "The Margarine Enquiry Commission," *Report* (1960) W.J. Waines to George Hutton.
48 "Speaker's Vote Shelves Bill to Color Manitoba Margarine," *Financial Post*, 2 April 1960, 20.
49 "Triangular Patties for Margarine, Squares for Butter in Manitoba," *Globe and Mail*, 12 May 1961, 1.
50 Manitoba, Legislative Assembly, *Journals, 1974*, 32 *passim*, and *Statutes of Manitoba, 1974*, 22-23 Eliz. II, cap. 10, 47.
51 "Co-op Will Make Margarine Here, Pool Parley Told," *Saskatoon Star Phoenix*, 17 November 1951, 3, and "Co-op and Dairy Pool Agree to Produce Substitute for Butter Here," ibid., 23 November 1951, 3.
52 C.H.P. Killick to V. McCormick, "Margarine File," PAM, Manitoba Department of Agriculture, 30 October 1959.
53 Douglas Papers, T.C. Douglas to J.S. Turnbull, 2 June 1955; T.S. Bentley to J.S. Turnbull, 26 July 1956.
54 Douglas Papers, T.C. Douglas to I.C. Nollet, 7 March 1958, copy; I.C. Nollet to T.C. Douglas, 1 April 1958.
55 Saskatchewan, Legislative Assembly, *Debates, 1965*, 418-19; *Journals, 1965*, 25 *passim*; and *Statutes, 1965*, 13-14 Eliz. II, cap. 10, 31.
56 Saskatchewan, *Debates, 1972*, 41 *passim*; *Journals, 1972*, 27 *passim*; and *Statutes, 1972*, 20-21 Eliz. II, cap. 73, 324.
57 Saskatchewan, *Debates, 1973-74*, 865-66; *Journals, 1973-74*, 47 *passim*, and *Statutes, 1973-4*, 22-23 Eliz. II, cap. 60, 254.
58 "Alberta Government Refuses to Allow Colored Oleo. Yellow Margarine Might Upset Balance, Ure States," Alberta Scrapbook Hansard, Reel 15, 18 March 1952.
59 "Colored Margarine Bill Progresses," ibid., Reel 20, 24 March 1964; Alberta, Legislative Assembly, *Journals, 1964*, vii *passim*; and *Statutes of the Province of Alberta, 1964*, 13 Eliz. II, cap. 51, 189-90.
60 *Alberta Hansard, 1972*, 30 *passim*; *Journals, 1972*, ii *passim*; *Statutes, 1972*, 21 Eliz. II, cap. 64, 240.
61 "Woman's Victory" (editorial), *Vancouver News-Herald*, 7 March 1952, 14; British Columbia, Legislative Assembly, *Journals, 1952*, 9 *passim*; and *Statutes, 1952*, 1 Eliz. II, cap. 9, 45-46.
62 British Columbia, Legislative Assembly, *Debates, 1973*, 358-59; *Journals, 1973*, 26 *passsim*; and *Statutes, 1973*, 22 Eliz. II, cap. 135, 247.
63 British Columbia, Legislative Assembly, *Debates, 1973*, 359.
64 CBC Commentary on station CBL, the day after the 1985 provincial election.
65 "Oleo Fight Not Over, Farm Group Indicates," *Globe and Mail*, 9 January 1953, 31.
66 Ontario, *Journals, 1953*, 32 *passim*, and *Statutes, 1953*, 1 Eliz. II, cap. 31, 259-60.
67 "Frost Flays Distortion by Edible Oils Groups, as Revised Bill Read," *Globe and Mail*, 25 March 1953, 9.
68 Ontario, *Journals, 1953*, 106 and 115, and "Technicality Blocks Margarine Showdown," *Globe and Mail*, 26 March 1953, 9.
69 "Creamerymen Seek Tax of Fifteen Cents on Oleo," ibid., 5 December 1953, 1.
70 "Ask Ban on Sale of Margarine by Deliverymen," ibid., 23 February 1962, 81.
71 Ontario, *Debates, 1962-63*, 2639 *passim*; *Journals, 1962-63*, 16 *passim*; and *Statutes, 1962-63*, 11-12 Eliz. II, cap. 93, 449-50.
72 "Here Comes Marge Bill Backed by 50,000 Buyers," *Financial Post*, 20 March 1965, 37.
73 Ontario, *Debates, 1962-63*, 2640.
74 Ontario, *Agricultural Statistics for Ontario, 1941-1978*, (1979) exposures 12 and 13.
75 "Margarine Bootlegging Reaching Fever Pitch," *London Free Press*, 11 October 1983, D9.

76 "Ontario May Let Margarine Firms Use the Same Yellow as Butter," *Record*, 28 September 1985, B15, and "Ontario Can't Stop Butter-Like Margarine," ibid., 22 August 1986, B11.
77 Bryan Stortz, Kitchener, to author, interview, 10 April 1987, and "An Act to Amend the Oleomargarine Act," *Statutes of Ontario, 1986*, 34-35 Eliz. II, cap. 65, 697.
78 Jim Romahn, "Satisfied Customers Call to Defend Super Centre," *Record*, 29 April 1987, D1.
79 "Margarine Will Be Darker," *Gazette* (Montreal), 18 April 1987, A4, and "Quebec Margarine Must Change Color," *Record*, 7 August 1987, B15.
80 "Quebec Margarine Firms Launch Offensive Against Change in Color," ibid., 11 August 1987, B7, and "The Facts About Butter" (full-page advertisement), *Gazette*, 15 August 1987, A-7.
81 P.E.I., Department of Agriculture, Tim Carrol to author, 20 August 1987.

Chapter 10

1 *Historical Statistics* (1983), series M310 and 342, and *Royal Commission on Canada's Economic Prospects, Final Report* (Ottawa, 1957), 169.
2 Jim Romahn, "Milk Pricing Madness,"*Kitchener-Waterloo Record*, 10 January 1979, 24.
3 *Canada's Economic Future* (Toronto: Cockfield Brown, 1956), 225-27.
4 David W. Slater, *Canada's Imports* (Study for the Royal Commission on Canada's Economic Prospects) (Ottawa, 1957), 180.
5 *Canada's Economic Future*, 225-27.
6 Ibid., 96-98.
7 *Historical Statistics* (1983), series A1, M342 and M368.
8 Ibid., series M357-62, M438 and M440; Statistics Canada, "Per Capita Food Disappearance, 1960-1982" (1982); and Statistics Canada, "Apparent Per Capita Food Consumption in Canada," Part I, annual 1983-87.
9 *Historical Statistics* (1983), series M156.
10 "Where the Farmer Stands" (editorial), *Guardian* (Charlottetown), 12 May 1947, 4.
11 *Historical Statistics* (1965), series L114-18.
12 John H. Young, *Canadian Commercial Policy*, Study for the Royal Commission on Canada's Economic Prospects (Ottawa, 1957), 168-72.
13 Don Mitchell, *Politics of Food* (Toronto: James Lorimer, 1975), 120-26.
14 Ibid., 115.
15 Canadian Facts Ltd., "Report of a Study of Consumer Attitudes Toward Margarine in Manitoba" (conducted for Manitoba Department of Agriculture and Immigration) (Toronto, 1953).
16 Canada, Royal Commission on Prices, *Report* (Ottawa, 1949), 45.
17 "National Milk Shortages a Conundrum," *Record*, 21 November 1983, B9, and Jim Romahn, "Food Views: Ontario Butter May Disappear," ibid., 11 December 1981, 31.
18 "Butter Industry Sees Death Knell in Milk in Allocation Cuts," ibid., 24 November 1979, 29.
19 *Historical Statistics* (1983), series M363-68.
20 Ibid., series M438 and M440; "Per Capita Food Disappearance, 1960-82"; and "Apparent Per Capita Food Consumption in Canada," Part I (1983-87).
21 Ibid., series M238.
22 *Urban Retail Food Prices 1914-1959* (Ottawa, 1960), 61.
23 *Historical Statistics* (1983), series K9.
24 Defence Research Medical Laboratories, *Development of an All-Temperature Canadian Biscuit Spread* (Ottawa, 1958).
25 Margaret Visser, *Much Depends on Dinner: The History and Mythology of a Meal* (Toronto: McClelland and Stewart, 1986), 104.
26 "Could Use Rapeseed," *Leader-Post* (Regina), 14 December 1948, 1.

27 "Can Use Seal Oil," *Globe and Mail*, 5 January 1951, 15.
28 T.C. Douglas Papers, H.S. Hanna to W.H. Horner, 6 January 1954.
29 "Rapeseed Becoming a Popular Export," *Record*, 10 September 1979, 27.
30 Jim Romahn, "Food Views: Rapeseed Experiments in Ontario," ibid., 31 August 1979, 22; *Saturday Night*, May 1980, 11; and Jim Romahn, "Food Views: Better Canola," *Record*, 19 October 1984, B11.
31 "Head-to-Head Race," ibid., 13 August 1985, B7.
32 John G. Kaffezakis, "Margarine: Canada," *Foreign Trade*, 8 August 1964, 11-12.
33 United Nations, *International Trade Statistics Yearbook, 1984* (New York, 1986), 67.
34 Canadian Edible Oil Food Industries, "Brief to Royal Commission on Canada's Economic Prospects" (1956), 26; "Soybean Growers Said Unhappy Over Compromise Edible Oils Bill," *Globe and Mail*, 27 March 1953, 5; T.C. Douglas Papers, H.L. Fowler to T.C. Douglas, 28 January 1954; and Stuijvenberg, *Margarine*, 25.
35 "New Butter Substitute Would Cost Much Less," *Calgary Herald*, 24 June 1948, 9.
36 "Ontario To Aid Search for Butter Substitutes," *Globe and Mail*, 17 February 1951, 1.
37 H.A. Hansen et al. "Flavour Stability of Canned Butter Concentrate," *Canadian Dairy and Ice Cream Journal* (September 1957): 53, 54 and 74.
38 T.C. Douglas Papers (copy) G. Unwin to J.S. Turnbull, 8 January 1952.
39 "Whipped Butter Called a Mystery to Consumers," *Record*, 21 November 1980, 27.
40 "High-price Spread: The Hunt Is Costly For a Better Butter," *Globe and Mail*, 22 May 1982, 17.
41 "Quebec Paves Way for Low-Calorie Butter," *Record*, 27 May 1980, 24, and Jim Romahn, "Food Views: Niagara Lard Isn't All Gold," ibid., 30 January 1984, B7.
42 Jim Romahn, "Creamery 'Revolution'," ibid., 21 April 1989, B11.
43 "Butter Lobby Digs in for Margarine Battle," ibid., 16 May 1978, 23.
44 "Researcher Uses Old Theory to Take Cholesterol Out of Butter," ibid., 1 October 1986, E1.
45 "Huge Vats of Oil Imperil the Dairy Industry," *Farmer's Advocate*, 28 June 1952, 5.
46 Ontario, Ministry of Agriculture and Food, "Report of Oleomargarine-Butter Blend Committee," January 1980, *Sessional Paper* No. 61, Tabled 17 April 1980.
47 "Protection for Butter Industry Cited in Ban on Spread Blends," *Record*, 22 April 1980, 28.
48 Jim Romahn, "Some Vets Abuse Powers, Butter Blends," ibid., 20 April 1979, 26.
49 *Toronto Sunday Star*, 9 August 1987, F1.
50 Royal Commission on Price Spreads of Food Products, *Report*, Vol. II (1959), 179; Institute of Edible Oil Foods, "Farm Gate Value of Soybean Oil," unpublished paper, 10 February 1982.
51 L.R. Rigaux, "A Preliminary Paper on the Canadian Edible Oils Industry" (Food Prices Review Board, *Reference Paper No. 4*), 1976, 67.
52 House of Commons, *Sessional Papers, 1954*, No. 204, not printed.
53 Bernd Weinberg, "Trends in Canadian Margarine Legislation," *Journal of the Canadian Institute of Food Science Technology* 7, 1 (January 1974): A12-A13.
54 *Scott's Directories* (Oakville, Ontario) for 1985-1986 and 1986-1987.
55 Rigaux, "A Preliminary Paper on the Canadian Edible Oils Industry," 69.
56 Jim Romahn, "Canada Fares Poorly on Milk Costs," *Record*, 16 January 1989, B5.

Chapter 11

1 John Downing, "Holy Cow, It's Still a Holy Cow," *Sunday Star* (Toronto), 17 January 1982, 22.
2 Mary Douglas, *Purity and Danger: An Analysis of Concepts of Pollution and Taboo* (New York: Praeger, 1970).

3 This argument is based primarily on Visser, *Much Depends on Dinner*, especially the chapter on butter, and her interview with Peter Gzowski, carried as a repeat on CBL on 30 June 1987.
4 James P. Johnston, *A Hundred Years of Eating: Food, Drink and the Daily Diet in Britain Since the Late Nineteenth Century* (Montreal: McGill-Queen's Press, 1977), 115.
5 L.B. Pett, "Are There Nutritional Problems in Canada?" *Canadian Medical Association Journal* 58, 11 (October 1948): 326-38.
6 "Opening Sessions *UNSCCUR* Underway at Lake Success," *Manitoba Cooperator*, 25 August 1949, 1.
7 L.B. Pett, "A New Dietary Standard for Canada, 1949," *Canadian Medical Association Journal*, 59, 12 (November 1949): 451, 455.
8 Zachary Sabry et al., "Nutrition Canada," *Nutrition Today* (January-February 1974): 1.
9 Food Prices Review Board, *What Price Nutrition?* (1975).
10 National Health and Welfare Canada, Food Directorate, *Recommended Nutrient Intakes for Canadians* (1983), Table X.1.
11 Agriculture Canada, Marketing and Economics Branch, *The Apparent Nutritive Value of Food Available for Consumption in Canada, 1960-75*, edited by Linda Robbins and Sushma Barewal (1981), Table J.1., 121 and Table K.1., 104, 128.
12 Papers of the Dairy Farmers of Canada, Joint Agricultural Producers Committee of Manitoba, "Brief Filed with the Margarine Inquiry Commission," Winnipeg, 27 October 1959.
13 "Myths and Fallacies About Good Nutrition," *Kitchener-Waterloo Record*, 7 August 1985, F1.
14 "Butter Is Supreme," *Manitoba Cooperator*, 15 January 1948, 11.
15 "Flouridated Milk and the Man Who Wants Our Children to Drink It," *Nutrition Quarterly*, 8, 1 (1984): 4-5.
16 Department of Industry, Food Products Branch, *Proceedings of the Symposium on Margarine and New Edible Oil Products*, presented at the Annual Conference of the Canadian Institute of Food Technology, 9-12 June 1968, (Ottawa 1968), 26.
17 P. Allen, "La Margarine," *L'Actualité Economique*, 30 (1954): 188-90.
18 *Record*, 28 May 1980, 16.
19 N.M. Kudelko, editorial, *Annals of Allergy*, April 1970, 164-65.
20 Jim Romahn, "Food Views," *Record*, 28 January 1981, 32.
21 Germain J. Brisson, et al., *Report of the Ad Hoc Committee on the Compositions of Special Margarines: A Critique* (Ottawa: Dairy Bureau of Canada, 1981).
22 "The Countrywoman," *Country Guide*, February 1955, 51.
23 A. McGill, "The Control of Dyes in Our Food Products," *Industrial Canada*, November 1923, 82.
24 "Coloured Margarine Plea Fades Away,"*Halifax Herald*, 16 February 1961, 1.
25 "Butter, Margarine Debate to Heat Up," *Record*, 4 December 1981, 27.
26 "Quebec Farm Group President Wants 'Deep Green' Margarine," *Globe and Mail*, 30 December 1948, 3.
27 "Farmers Urge Brown Margarine," *Toronto Star*, 13 February 1962, n.p.
28 "Displaying the Colours" (editorial), *Record*, 1 February 1982, 18.
29 "A Fat Lot of Good" (letter to the editor), *The Times* (London), 29 May 1982.
30 "Women's Victory" (editorial), *Vancouver News Herald*, 7 March 1952, 4.
31 Papers of Dairy Farmers of Canada, letter from Ontario Federation of Agriculture and Cream Producers of Ontario to Premier of Ontario, 25 February 1949, copy.
32 "Dairymen Deserve This Protection" (editorial), *Montreal Daily Star*, 16 March 1949, 12.
33 Papers of the Dairy Farmers of Canada, "The Need for a Distinctive Margarine Colour," July 1981.
34 Ibid.
35 Ibid.
36 David Scott-Atkinson, "I've Hardly Ever Met an American I Didn't Like...Almost," *Mississauga News*, 18 March, 1981, n.p.

37 Papers of the Dairy Farmers of Canada, K.A. Standing, Secretary-Manager, Ontario Soya Bean Growers Marketing Board, Chatham, Ontario to Presidents and Directors, Institute of Edible Oil Foods, 6 February 1960 (copy).
38 "Butter, Margarine Debates to Heat Up," *Record*, 4 December 1981, 27.
39 "Color of Margarine Worth $22 Million, Farmers Told," ibid., 14 January 1988, B9.
40 Institute of Edible Oil Foods, "Summary of Position re Margarine Colour Restrictions in Ontario, Recommendation to the Ontario Government," 7 December 1981.
41 Institute of Edible Oil Foods, "Proposal to the Hon. William Newman, Minister of Agriculture and Food, Ontario Provincial Government, for the Removal of Colour Restrictions on Margarine" (October 1976).
42 "All Stirred Up Over Margarine," *Globe and Mail*, 26 January 1961, 4, and "Margarine Fight to Continue," ibid., 13 January 1962, 11.

Chapter 12

1 The essence of this chapter is based on material from the excellent essay written in 1980-81 by Bryan Stortz in History 680 at Wilfrid Laurier University: "A History of the Struggle Between Butter and Margarine for the Canadian Market."
2 "Ban on Margarine Ingredients Proposed at Dairy Council," *Globe and Mail*, 22 March 1940, 2.
3 Daniel Melnick, "Development of Organoleptically and Nutritionally Improved Margarine Products," *Journal of Home Economics* (December 1968): 794.
4 S.F. Riepma, *The Story of Margarine* (Washington, D.C.: McGraw-Hill Ryerson, 1970), 91.
5 Stuijvenberg, *Margarine*, 536.
6 "Margarine Sells Here Saturday," *Vancouver Daily Province*, 17 December 1948, 1.
7 "Nothing Is Certain," *Calgary Herald*, 14 December 1948, 1, and Massey to Blackwell, 20 December 1948, PAO, R.G. 3, Department of Agriculture Records.
8 *Canadian Home Journal*, June 1949, 38.
9 "First Canadian Made Oleo Sells in B.C. for 53 Cents," *Kitchener-Waterloo Record*, 21 December 1948, 17.
10 *Canadian Home Journal*, September 1949, 4.
11 Ibid., February 1941, 2.
12 Ibid., March 1951, 2.
13 Ibid., November 1952, 38.
14 Melnick, "Development of Organoleptically and Nutritionally Improved Margarine Products," 795.
15 Ibid., 796; *Canadian Home Journal*, February 1960, 14.
16 "Facts About Margarine" (editorial), *Brantford Expositor*, 18 December 1948, 4.
17 Euler Papers, H.B. Wright, Calgary, to Euler, 8 February 1949.
18 *Historical Statistics* (1965), series L 108.
19 "Oleo on Sale in Vancouver, Snapped Up at 53 Cents per Pound," *Globe and Mail*, 21 December 1948, 1.
20 *Record*, 14 September 1949, 10.
21 "Preview of Margarine Given Dietitians, Press," *Globe and Mail*, 24 December 1948, 1.
22 "Watch the Butter Lobby," *Maclean's*, 1 February 1949, 2, and C. Fred Bodsworth, "Bread's Other Spread," ibid., 15.
23 "We Can Make These With Butter or Margarine," *Canadian Home Journal*, March 1949, 34.
24 Interview with Allan J. Phillips of Monarch Fine Foods, Toronto by Bryan Stortz, 9 November 1981, and William Applebaum, *Brand Strategy in United States Food Marketing* (Boston: Harvard University Press, 1957), 17.
25 Phillips/Stortz interview.

26 Institute of Edible Oil Foods, *Submission to the Ontario Government for the Removal of Colour Restrictions on Margarine* (Toronto, 1976), 5.
27 Wilson, *History of Unilever*, III, 248.
28 "Calls for Fight on Coloured Oleo," *Globe and Mail*, 25 November 1954, 5, and "Dairy Industry's Margarine Stand Said Incredible," ibid., 22 November 1955, 2.
29 "Housewife's Stand vs. the Cow's" (editorial), *Vancouver News-Herald*, 13 May 1949, 7.
30 *Canadian Home Journal*, June 1949, 38.
31 *Globe and Mail*, 12 July 1950, 4.
32 *Canadian Home Journal*, May 1950, 76. The italics are in the original advertisement.
33 Ibid., March 1941, inside front cover.
34 Ibid., February 1951, 2.
35 Ibid., December 1954, 26.
36 Contest forms are in PANB New Brunswick, Department of Agriculture, RAG 1953, RS 124, Box 13, Dairying, Miscellaneous 1953.
37 Ibid.
38 *Globe and Mail*, 8 October 1951, 4.
39 *Canadian Home Journal*, September 1950, 78.
40 "Fight for Your Product, Pro-Butter Forces Told: 'I like Oleo,' Says One," *Globe and Mail*, 5 January 1951, 15.
41 T.C. Douglas Papers, copy, J.S. McLean, President, Canada Packers, to T.L. Kennedy, Premier of Ontario, 22 February 1949.
42 "Margarine Output Tops Butter First Time in Canada's History," *Globe and Mail*, 13 March 1950, 23.
43 *Historical Statistics* (1983), series M344 and M355.
44 *Chatelaine*, February 1960, 14; ibid., April 1967, 12.
45 "Dairy Official Claims Margarine Savings Slight," *Winnipeg Free Press*, 15 December 1948, 9.
46 "Margarine Brings Depression to Canada, Dairyman Claims," *Edmonton Bulletin*, 20 March 1950, 5.
47 "Hannam Says Canadian Farmers Seeking Reasonable Price Levels," *Manitoba Cooperator*, 23 February 1950, 1.
48 "The Anti-Butter Brigade Abandons the Poor" (editorial), *Farm and Ranch Review*, April 1949, 6.
49 "Request Issued Not to Buy Oleo," *Calgary Herald*, 16 October 1950, 1.
50 "Farmers Heavy Buyers of Oleo," *Leader Post* (Regina), 1 April 1950, 2.
51 "Small Dairymen Large Margarine Consumers," *Red Deer Advocate*, 28 June 1950, section 2, 7.
52 Interview with Dr. Gerald Stortz, Guelph, Ontario by author, 21 May 1981, and interview with Mrs. Fran Bowden, Charlottetown, P.E.I. by author, 8 October 1980.
53 C.O. Ilori, "Analysis of Retail Store Merchandising Practices of and Consumer Preferences for, Butter and Margarine," Master's Thesis, University of Toronto, 1958, 18, and *Canadian Consumer Survey* (Toronto, 1965), 36.
54 Ibid., 19 and 20.
55 "Dairymen Fear Price Break if Imports Not Controlled," *Globe and Mail*, 18 March 1952, 1.
56 "Margarine, For Love or Money" (editorial), *Maritime Cooperator*, 15 October 1948, 8.
57 "Margarine Has Friend," *Leader-Post* (Regina), 26 January 1949, 2.
58 *Farm Forum Guide*, 8 January 1951, 3.
59 "Canadian Dairy Farmers Plan Two-Fold Program," *Manitoba Cooperator*, 2 February 1950, 1.
60 Bernd Weinberg, "Trends in Canadian Margarine Legislation," A12-A14.
61 *Farm Forum Guide*, 8 January 1951, 2, and *Canadian Home Journal*, November 1951, 28.
62 *Consumer Report*, May 1963, 2, 34.
63 Germain J. Brisson, *Lipids in Human Nutrition, An Appraisal of Some Dietary Concepts* (Englewood, N.J.: Jack K. Burgess, 1981), chap. 4.

64 David Scott-Atkinson, "A Summary Report on the Public Relations Program of the Institute of Edible Oil Foods," report 2, 21 October 1969.
65 Interview with David Scott-Atkinson by author in Toronto, 21 June 1980.
66 *Chatelaine*, January 1979, 60.
67 Ibid., January 1982, 79.
68 "Margarine Gifts for Doctors 'Unethical' But Not Unlawful," *Globe and Mail*, 31 July 1980, 10.
69 "Margarine Has Edge as Market Struggle with Dairymen Looms," *Record*, 27 December 1977, 35.
70 *Chatelaine*, September 1979, 141.
71 *Record*, 27 June 1979, 12.
72 *Chatelaine*, December 1979, 140.
73 Ibid., October 1979, 167.
74 Ibid., August 1979, 82.
75 Ibid., February 1982, 183.
76 "Milk Board Fears Margarine-blend Competition," *Record*, 22 June 1979, 28.
77 Ibid., 17 September 1980, 20.
78 C.G. Sheppard, Consumer and Corporate Affairs Canada, Hull, Quebec, to author, 12 May 1987.
79 Andrew Cohen, "What's Healthy, What's Not? Butter, Margarine Makers Escalate the Argument," *Financial Post*, 9 March 1985, 25.
80 "Margarine Producers Protest Dairy Ads," *Record*, 17 September 1980, 35.
81 "Federal Campaigns Take Opposite Stands on Butter Consumption," ibid., 10 October 1978, 4, and "Conflicting Ads on Butter Noted," ibid., 12 October 1978, 31.
82 Dairy Bureau of Canada, "The Rationale and Need for Industrial Milk Promotion and a Funding Recommendation for 1982", Toronto, 13 and 14 January 1982, 17.
83 Ibid., 18.
84 *Record*, 13 January 1984, 37.
85 *Macleans*, 12 April 1982, 22.
86 *Chatelaine*, April 1974.
87 Ibid., February 1982, 78.
88 *Record*, 3 June 1980, 5.
89 Ibid., 28 January 1981, 37.
90 "Report on Market Share," *Financial Times of Canada*, 22 September 1985, 33.
91 Ibid., 31 October, A8.
92 "Food Views: Farmers Love Television Ads," *Record*, 14 February 1982, 17.
93 Interview with Jim Trouble, Dairy Bureau of Canada, Toronto, by Bryan Stortz, 12 January 1982.

Conclusion

1 H.J. Teuteberg, "The General Relationship Between Diet and Industrialization," in Elborg and Robert Forster (eds.), *European Diet from Pre-Industrial to Modern Times* (New York: Harper & Row, 1975), 71.
2 Ibid., 67.
3 Ibid., 72-73.
4 Roland Barthes, "Toward a Psychosociology of Contemporary Food Consumption" in Forster, *European Diet*, 49.
5 Visser, *Much Depends on Dinner*, 105.
6 Ibid., 103-105.

BIBLIOGRAPHY

Abbreviations

BCA — British Columbia Archives
NAC — National Archives of Canada
PAA — Provincial Archives of Alberta
PAM — Provincial Archives of Manitoba
PANB — Provincial Archives of New Brunswick
PANS — Provincial Archives of Nova Scotia
PAO — Provincial Archives of Ontario
QUA — Queen's University Archives
SAB — Saskatchewan Archives Board

Government Documents and Archival Holdings

Federal

Agriculture Canada. Marketing and Economics Branch. *Apparent Nutritive Value of Food Available for Consumption in Canada 1960-1975.* Edited by Linda Robbins and Sushma Barewal. 1981.
———. *Handbook of Food Expenditures, Prices and Consumption.* 1977-84.
Canada Gazette, 1886-1988.
Canada Law Reports. 1949.
Canada Year Book, 1905-1987.
Census of Canada, 1871-1981.
Defence Research Medical Laboratories. *Development of an All-Temperature Canadian Biscuit Spread.* Ottawa: Defence Research Board, 1958.
Department of Agriculture. RG17, BIII. Dairy Products Division. Vol. 2540, File D10, "Butter Substitutes, Oleomargarine, Butter General." Department of Health. *Food and Drug Regulations.*
Department of Industry, Food Products Branch. *Proceedings of the Symposium on Margarine and New Edible Oil Products.* Presented at the Annual Conference of the Canadian Institute of Food Technology, June 9-12, 1968.
Dominion Bureau of Statistics. Industry & Merchandising Division, Animal Products Section. *Margarine Statistics.*
Drummond, W.M. and W. Mackenzie. *Progress and Prospects of Canadian Agriculture* (Study for Royal Commission on Canadian Economic Prospects). 1957.
Food Prices Review Board. *A Preliminary Paper on Some Food Policy Aspects of Nutrition and Health* (Wirick Report). 1976.
———. *What Price Nutrition?* 1975.
Health and Welfare Canada. *Recommended Nutrient Intake for Canadians.* 1983.
House of Commons. *Debates, 1867-1989.*
———. *Journals, 1867-1989.*
———. *Sessional Papers and Annual Departmental Reports.* 1867-1929.
Pett, Dr. L.B. (compiler). *Table of Food Values Recommended for Use in Canada.* Department of National Health and Welfare, Nutrition Division, 1946.
Proceedings of the Special Joint Commission on Consumer Credit. House of Commons, 1964-67.
Report of the Ad Hoc Committee on Composition of Special Margarines. Ottawa, 1979 (1980).

Rigaux, L.R. "A Preliminary Paper on the Canadian Edible Oils Industry," Reference Paper No. 4. Food Prices Review Board, 1976.
Royal Commission on Prices. *Report*. 1949.
Royal Commission on Canada's Economic Prospects. *Report*. 1956.
Royal Commission on Price Spreads of Food Products. *Report*. 1960.
Royal Commission on Taxation. *Report*. 1967.
Saunders, S.A. *The Economic History of the Maritime Provinces* (Study for the Royal Commission on Dominion-Provincial Relations). 1939.
Senate. *Debates, 1867-1989.*
———. *Journals, 1867-1989.*
Slater, David W. *Canada's Imports* (Study for the Royal Commission on Canada's Economic Prospects). 1957.
Statistics Canada. "Apparent Per Capita Food Consumption in Canada," Part I. 1980-87.
———. *Cereals and Oilseeds Review*. 1978-87.
———. *Family Food Expenditure*. Various years.
———. *Handbook of Agricultural Statistics*, Part VII *Dairy Statistics 1920-73*. 1975.
———. *Oils and Fats*. 1950-87.
———. "Per Capita Food Disappearance, 1960-1982."
———. *Urban Retail Food Prices, 1914-59*. 1960.
Statutes of Canada. 1867-1987.
Supreme Court Reports. 1949.
Young, John H. *Canadian Commercial Policy* (Study for the Royal Commission on Canada's Economic Prospects). 1957.

Provincial Government Documents

Also in Provincial Archives

Alberta

Department of Agriculture. *Annual Report*. 1905-65.
———. Dairy Branch Files, 1904-52.
———. Ministers and Deputy Ministers files, c. 1939-70.
Federation of Agriculture. Dairying (Pegging of Butter Prices), 1940-70.
Legislative Assembly. *Hansard*, 1972.
———. *Journals*, 1921-79.
Premiers' Papers. File 1529, Agricultural Products, 1949-55.
Scrapbook Hansard. 1949-64.
Statutes of Province of Alberta. 1949-72.

British Columbia

Department of Agriculture. Dairy Branch, "Minute Books, etc. of Dairymen's Association, 1894-1932."
———. *Report*. 1895-1974.
Legislative Assembly. *Journals*. 1917-79.
———. *Debates*. 1973.

———. *Sessional Papers.* 1872-1980.
Shuswap Okanagan Dairy Industries Cooperative Association, "Brief re Margarine and Its Effect on the Dairy Industry." 1949.
Statutes. Revised, 1948.
Vancouver Junior Chamber of Commerce. Views on Questions of Public Concern. Addressed to Premier and Members of the Cabinet, Victoria, 1949.

Manitoba

Department of Agriculture. *Annual Report.* 1917-52.
———. *Report* of study (by Canadian Facts Ltd.) "Consumer Attitudes towards Margarine in Manitoba." Toronto, 1953.
———. *Proceedings* of First Meeting of Wartime Committee on Agriculture, 1940.
Legislative Assembly. *Debates and Proceedings.* 1958-61.
———. *Journals.* 1949-79.
Margarine Enquiry Commission. *Report.* 1960.
Statutes of Manitoba. Revised, 1970.
———. 1949.

New Brunswick

Acts of the Legislature. 1935-80.
Department of Agriculture. "Dairying." 1942-53.
Department of Health. *Report.* 1970-71.
Legislative Assembly. *Journals.* 1886-1977.
———. "Trade and Navigation Returns for the Province of New Brunswick." 1868.
———. "Report on Agriculture." 1867-86.
Report on Agriculture for the Province of New Brunswick. 1917-60.
Statutes of New Brunswick. Revised, 1952.
Synoptic Report of the Proceedings of the Legislative Assembly of the Province of New Brunswick. 1949-77.

Newfoundland

Bridle, Paul (ed.). *Documents on Relations Between Canada and Newfoundland.* Ottawa, 1984.
Census of Newfoundland and Labrador. 1857-1911.
Cuthbertson, D.P. *Report on Nutrition in Newfoundland.* London, 1947.
Department of Public Health and Welfare. *Annual Report* of Medical Services Division. 1946.
Directory for Towns of St. John's, Harbour Grace, and Carbonear, 1885-86. 1885.
Food and Drug (Margarine) Regulations. 1966.
House of Assembly. *Verbatim Report.* 1973.
McAlpine Newfoundland Directory. 1898.
Might and Company's Directory, St. John's, Harbour Grace and Carbonear. 1890.
National Convention. *Proceedings.* 1946-48.
Newfoundland Directory. St. John's, 1928, 1936.

Report and Documents related to Negotiations for Union of Newfoundland with Canada. 1949.
St. John's Classified Business and City Directory 1932. 1932.

Nova Scotia

Department of Agriculture. *Annual Report.* 1956-64.
House of Assembly. *Debates.* 1951-1974.
House of Assembly. *Journals and Proceedings of the House of Assembly.* 1920-80.
Legislative Council. *Debates.* 1881-88.
Statutes. Revised, 1949, 1954, 1967.

Ontario

Agricultural Statistics for Ontario 1941-1978. 1979.
Department of Agriculture. *Annual Report.* 1886-1979.
Department of Agriculture Papers.
Farm Economics and Statistics Branch. Study: "Consumer Purchases and Opinions of Butter and Margarine in One Supermarket in Newmarket, August 1961."
Legislative Assembly. *Sessional Papers.* 1869-1948.
Legislature of Ontario. *Debates.* 1947-87.
_____. *Journals.* 1917-23.
Ministry of Agriculture and Food. "Report of Oleomargarine-Butter Blend Committee." January 1980.
Ontario Agricultural Commission. *Report*, plus appendices. 1881.
Ontario Royal Commission on Milk. *Report.* 1947.
Ontario Royal Commission on Price Spreads of Food Products. *Submission.* September 15, 1958.
Statutes. 1949.

Prince Edward Island

General Assembly. *Acts.* 1965-80.
Legislative Assembly. *Journal.* 1867-1982.

Quebec

Department of Agriculture. *Rapport annuel [du] mérite agricole.* 1970-76.
Gazette Officielle du Québec. 1972.
"General Report of Commissioner of Agriculture and Public Works." *Sessional Papers.* 1885.
_____. J.C. Chapais, "The Past, the Present and the Future of the Dairy Industry in the Province of Quebec." 1909.
Legislative Assembly. *Debates.* 1879-1968.
National Assembly. *Votes and Proceedings.* 1972-87.
_____. *Journals.* 1960-71.
Statutes of Quebec. Revised, 1964.

Saskatchewan

Department of Agriculture. *Annual Report of Dairy Commissioner.* 1913-51.
──────. Dairy Branch. Newsletter, "Dairymen Topics" (monthly).
──────. Dairy Branch. "Oleomargarine 1916-48," Correspondence of Secretary with Dairymen and Government Officials re Ban.
Legislative Assembly. *Debates and Proceedings.* 1947-74.
──────. *Journals.* 1917-79.
──────. *Sessional Papers.* 1964.
Royal Commission on Agriculture and Rural Life. *Report #10, The Home and Family in Rural Saskatchewan.* 1956.
Statutes.

International

United Nations. Food and Agriculture Organization. *FAO Trade Yearbook.* Vol. 13, 1959 and vol. 38, 1985.
──────. *International Trade Statistics Yearbook 1984.* Vol. 2: *Trade by Commodity.* New York, 1986.

Papers

British Columbia Federation of Agriculture, BCA.
Robert Borden Papers, NAC.
Stewart Cameron Collection, PAA.
Canadian Council of Agriculture, PAM.
Canadian Federation of Agriculture, NAC.
Consumers' Association of Canada, NAC.
Thomas L. Crerar Papers, QUA.
Dairy Farmers of Canada Archives, Ottawa.
T.C. Douglas Papers, SAB.
William Daum Euler Papers, NAC.
William C. Good Papers, NAC.
Institute of Edible Oils Archives, Toronto.
T.L. Kennedy papers, PAO.
Ian Mackenzie Papers, NAC.
William Lyon Mackenzie King Papers; also the Diaries, NAC.
Manitoba Provincial Council of Women, PAM.
Arthur Meighen Papers, NAC.
Montreal Council of Women, NAC.
Ontario Milk Marketing Board Papers, Toronto.
Sussex Cheese and Butter Company, PANB.
United Farmers of Alberta, Annual Report and Year Book, PAA.
United Farmers of Manitoba, PAM.

Newspapers

Acadien (Moncton), 1913-26.
Albertan (Calgary), 1947-51.
Brantford Expositor, 1943, 1948.
Calgary Daily Herald, 1919, 1946-61.
Canadian Gleaner (Huntingdon, Quebec), 1863-1900.
Citizen (Ottawa), 1917.
Courier de St. Hyacinthe, 1886.
Daily Colonist (Victoria), 1946-51.
Daily Examiner (Charlottetown), 1886.
Daily Gleaner (Fredericton), 1948-54.
Daily News (St. John's, Newfoundland), 1950-53.
Dawson Weekly News, 1946-50.
Digby Weekly Courier, 1917.
Edmonton Bulletin, 1886.
Edmonton Journal, 1946-50.
Evening Telegram (St. John's, Newfoundland), 1949, 1960.
Examiner (Charlottetown), 1886.
Financial Post, 1917, 1948-71.
Financial Times of Canada, 1986.
Free Press (Ottawa), 1886.
Globe (Toronto), 1886-1940.
Globe and Mail (Toronto), 1946-87.
Guardian (Charlottetown), 1917, 1937-74.
Halifax Chronicle, 1946-48.
Halifax Chronicle-Herald, 1949-74.
Halifax Herald, 1947-48.
Kitchener Daily Telegraph, 1917-18.
Kitchener News Record, 1917-18.
Kitchener-Waterloo Record, 1920-90.
Le Devoir (Montreal), 1917.
Leader-Post (Regina), 1947-52.
Lethbridge Herald, 1946-61.
Lethbridge News, 1886.
London Advertiser, 1917.
London Free Press, 1980-87.
Mail (Brandon), 1886.
Manitoba Cooperator, 1947-50.
Manitoba Free Press, 1917.
Maritime Cooperator, 1946-50.
Maritime Merchant and Commercial Review, 1917.
Mississauga News, 1981.
Moncton Daily Times, 1886, 1946-73.
Monetary Times, 1867-1980.
Montreal Daily Star, 1917 and 1949.
Montreal Gazette, 1921, 1985 and 1987.
Moose Jaw Times, 1950-54.
Morning Bulletin (Edmonton), 1917.

Morning Chronicle (Halifax), 1886.
Morning Star (Vancouver), 1927.
New Brunswick Reporter, 1886.
New York Times, 1946-70.
News Record (Berlin), 1893-1919.
Nouveau Monde (Montreal), 1885-86.
Observer (Sarnia), 1886.
Ottawa Daily Citizen, 1886, 1926.
Ottawa Evening Journal, 1917.
Ottawa Journal, 1926.
Patriot (Charlottetown), 1886.
Pionnier (Sherbrooke), 1886.
Prince Albert Times and Saskatchewan Review, 1886.
Qu'Appelle Progress, 1886.
Quebec Chronicle Telegraph, 1946-50.
Red Deer Advocate, 1946-50.
Saint John Globe, 1917.
Saskatchewan Herald, 1886.
Saskatoon Star-Phoenix, 1949-53.
Stanstead Journal, 1886.
Sunday Star (Toronto), 1982.
Telegraph-Journal (Saint John, New Brunswick), 1948-54.
The Times (London), 1982.
Toronto Daily Mail and Empire, 1886.
Toronto Star, 1948-70.
Toronto World, 1886.
Union des Cantons de L'Est (Arthabaskaville, Quebec), 1886.
Vancouver Daily Province, 1917, 1946-50.
Vancouver News-Herald, 1944-52.
Vancouver Sun, 1946-50.
Victoria Daily Times, 1917-67.
Wawanesa Optimist (Manitoba), 1946-50.
Windsor Record, 1917.
Winnipeg Free Press, 1886, 1946-50.
Winnipeg Tribune, 1949-52.

Journals

Annals of Allergy, 1970.
Canadian Consumer Survey, 1965.
Canadian Countryman, 1914-51.
Canadian Dairy and Ice Cream Journal, 1923-50.
Canadian Forum, 1947-87.
Canadian Grocer, 1923-60.
Canadian Home Journal, 1945-58.
Canadian Journal of Public Health, 1949.
Chatelaine, The Canadian Women's Magazine, April 1948.
Consumer Report, 1979.
Country Guide, 1947-48, 1955.
Family Herald and Weekly Star, 1947-52.

Farm and Ranch Review, 1946-50.
Farm Forum Guide, 1946-51.
Farmer's Advocate, 1867-1950.
Food for Thought, 1946-47, 1951-52.
Grain and Oilseeds Review, 1981.
Industrial Canada, 1923-24.
Maclean's, 1949, 1951.
Maritime Farmer and Cooperative Dairyman, 1895-1979.
Maritime Merchant and Commercial Review, 1918.
Nutrition Quarterly, 1980-87.
Ottawa Farm Journal, 1917.
Saturday Night, 1980.
Time Magazine, 1948.

Memoirs

Collins, Robert. *Butter Down the Well, Reflections of a Canadian Childhood*. Saskatoon: Western Producer, 1980.

Dancocks, D.G. *In Enemy Hands: Canadian Prisoners of War, 1939-45*. Edmonton: Hurtig, 1983.

Fraser, Donald. *The Journal of Private Fraser, 1914-1918: Canadian Expeditionary Force*. Victoria, B.C.: Sono Nis Press, 1985.

Grogen, John Patrick. *Dieppe and Beyond for a Dollar and a Half a Day*. Renfrew: Juniper, 1982.

Hibbert, Joyce. *Fragments of War: Stories from Survivors of World War II*. Toronto: Dundurn Press, 1985.

Lawrence, Hal. *A Bloody War: One Man's Memories of the Canadian Navy, 1939-45*. Toronto, 198?.

McClung, Nellie. *Clearing in the West*. Toronto: Thomas Allen, 1935.

Minifie, James M. *Homesteader: A Prairie Boyhood Recalled*. Toronto: Macmillan, 1972.

Mowat, Farley. *And No Birds Sang*. Toronto: McClelland and Stewart, 1979.

Prouse, Robert A. *Ticket to Hell via Dieppe: From a Prisoner's Wartime Log, 1942-1945*. Toronto: Van Nostrand Reinhold, 1982.

Reid, Gordon (ed.). *Poor Bloody Murder: Personal Memoirs of the First World War*. Oakville, Ont.: Mosaic Press, 1980.

Monographs

Abell, Albert. *Laskin's Canadian Constitutional Law*. Toronto: Carswell, 1975.

Agricultural Historians. *History and Development of the Dairy Industry in Prince Edward Island*. Charlottetown: New Horizons Program, 1978.

Allen, Patrick. *La Margarine, Peut-elle Remplacer le Buerre?* Montreal, 1955.

Anderson, A.J.C. and P.N. Williams. *Margarine*. 2nd ed., rev. Oxford: Pergamon Press, 1965.

Applebaum, William. *Brand Strategy in United States Food Marketing: Perspective on Food Manufacturers' and Distributors' Brands in the United States*. Boston: Harvard University Press, 1967.

Armstrong, Robert. *Structure and Change: An Economic History of Quebec.* Toronto: Gage, 1984.
Auld, Francis H. *Canadian Agriculture and World War II: A History of the Wartime Activities of the Canada Department of Agriculture.* Ottawa: Department of Agriculture, 1953.
Bailey, A.E. *Bailey's Industrial Oil and Fat Products.* New York: John Wiley, 1964.
Barss, Beulah M. *The Pioneer Cook: A Historical View of Canadian Prairie Food.* Calgary: Detselig, 1980.
Berton, Pierre. *Centennial Food Guide: A Century of Good Eating. An Anthology of Writings About Food and Drink.* Toronto: Canadian Centennial Publishing Co., 1966.
Bliss, J. Michael. *A Canadian Millionaire: The Life and Business Times of Sir Joseph Flavelle, Bart., 1858-1939.* Toronto: Macmillan, 1978.
Bothwell, Robert, Ian Drummond and John English. *Canada Since 1945: Power, Politics, and Provincialism.* Toronto: University of Toronto Press, 1981.
Brisson, Germain J. *Lipids in Human Nutrition, An Appraisal of Some Dietary Concepts.* Englewood, N.J.: Jack K. Burgess, 1981.
────── et al. *Report of the Ad Hoc Committee on the Composition of Special Margarines: A Critique, June 1981.* Ottawa: Dairy Bureau of Canada, 1981.
Britnell, G.E. and V.C. Fowke. *Canadian Agriculture in War and Peace, 1935-50.* Stanford, Calif.: Stanford University Press, 1962.
Brown, Robert Craig and Ramsay Cook. *Canada, 1896-1921: A Nation Transformed.* Toronto: McClelland and Stewart, 1974.
Canada's Economic Future: Digest of One Hundred and Twenty-five Submissions to the Royal Commission on Canada's Economic Prospects. Toronto: Cockfield Brown, 1956.
Canadian Annual Digest, 54th issue, 1949, *Constitutional Law II.*
Canadian Federation of Agriculture. *The Margarine Question!* Ottawa: Canadian Federation of Agriculture, 1947.
Canadian Institute of Food Technology. *Symposium on Margarine and New Edible Oil Products.* Ottawa, 1968.
Carrigan, Owen (compiler). *Canadian Party Platforms, 1867-1968.* Toronto: Copp Clark, 1968.
Chadwick, St. John. *Newfoundland, Island Into Province.* Cambridge: Cambridge University Press, 1967.
Christiansen, R.P. *Using Resources to Meet Food Needs.* Washington, 1948.
Church, Gordon C. *An Unfailing Faith: A History of the Saskatchewan Dairy Industry.* [Microform] Toronto: Micromedia, 1980.
Clark, Colin. *Starvation or Plenty?* New York: Taplinger, 1970.
Clayton, William. *Margarine.* London: Longmans, 1920.
Coughlin, Bing. *Herbie.* Toronto: Thomas Nelson, n.d.
Creighton, D.G. *John A. Macdonald.* Vol. 2: *The Old Chieftain.* Toronto: Macmillan, 1955.
Darcovich, W. *Rapeseed Potential in Western Canada: An Evaluation of a Research and Development Program.* Ottawa: Agriculture Canada, 1973.
Darmon, Rene, Michel Laroche and John Petrol. *Canadian Marketing: Principles and Applications.* Toronto: McGraw-Hill Ryerson, 1981.
Davidson, A.L. *The Genesis and Growth of Food and Drug Administration in Canada.* Ottawa: Department of National Health and Welfare, 1950.
Dawson, R. MacGregor. *William Lyon MacKenzie King, A Political Biography.* Vol. 1: *1874-1923.* Toronto: University of Toronto Press, 1958.
Devine, P.K. *Ye Olde St. John's.* St. John's, 1936.

BIBLIOGRAPHY

Douglas, Mary. *Purity and Danger: An Analysis of Concepts of Pollution and Taboo*. New York: Praeger, 1970.

Drummond, Ian. *Progress Without Planning: The Economic History of Ontario from Confederation to the Second World War*. Toronto: University of Toronto Press, 1987.

Drummond, J.C. and A. Wilbraham. *The Englishman's Food: A History of Five Centuries of English Diet*. London: Jonathan Cape, 1969.

Drummond, W.M., W.J. Anderson and T.C. Kerr. *A Review of Agricultural Policy in Canada*. Ottawa: Agriculture Economics Research Council of Canada, 1966.

Duguid, A. Fortescue. *Official History of the Canadian Forces in the Great War 1914-1919*. Vol. 1. Ottawa: King's Printer, 1938.

Fairbairn, Garry L. *Will the Bounty End? The Uncertain Future of Canada's Food Supply*. Saskatoon: Western Producer, 1984.

Farmilo, A. *Oleomargarine and the Physical Deterioration of the Working People*. Edmonton, 1924.

Feasby, W. R. (ed.). *Official History of the Canadian Medical Services 1939-1945*. 2 vols. Ottawa: Queen's Printer, 1953.

Fieldhouse, Paul. *Food & Nutrition: Customs & Culture*. London: Croom, Helm, 1986.

Forester, C.S. *The General*. Boston: Little, Brown and Co., 1947.

Forster, Elborg and Robert (eds.). *European Diet from Pre-industrial to Modern Times*. New York: Harper & Row, 1975.

Fowke, Vernon. *Canadian Agricultural Policy*. Toronto: University of Toronto Press, 1947.

Francis, R.D. and D.B. Smith. *Readings in Canadian History: Post-Confederation*. 2nd ed. Toronto: Holt, Rinehart and Winston, 1986.

George, Roy E. *A Leader and a Laggard: Manufacturing Industry in Nova Scotia, Quebec and Ontario*. Toronto: University of Toronto Press, 1970.

George, Susan. *How the Other Half Dies: The Real Reasons for World Hunger*. Montclair, N.J.: Allanheld Osman, 1977.

Goodwyn, Lawrence. *Democratic Promise: The Populist Movement in America*. New York: Oxford University Press, 1976.

———. *The Populist Moment*. New York: Oxford University Press, 1978.

Guest, Dennis. *Emergence of Social Security in Canada*. 2nd ed., rev. Vancouver: University of British Columbia Press, 1985.

Guillet, Edwin C. *Pioneer Farmer and Backwoodsman*. Toronto: Ontario Publishing Co., 1963.

Gussow, Jean. *The Feeding Web: Issues in Nutritional Ecology*. Palo Alto, Calif.: Bull Publishing, 1978.

Guthrie, Edward Sewall. *The Book of Butter: A Text on the Nature, Manufacture and Marketing of the Product*. New York: Macmillan, 1918.

Gwyn, Richard. *Smallwood, the Unlikely Revolutionary*. Toronto: McClelland and Stewart, 1968.

Hays, Samuel P. *Response to Industrialism, 1885-1914*. Chicago: University of Chicago Press, 1957.

Healthful Eating (National Health Series, No. 109). Ottawa: King's Printer, 1944.

Hibbert, Joyce (ed.). *The War Brides*. Toronto: PMA Books, 1978.

Hogg, Peter W. *Constitutional Law of Canada*. Toronto: Carswell, 1971.

Hooker, Richard. *Food and Drink in America: A History*. Indianapolis: Bobbs and Morrow, 1981.

Hunter, Beatrice T. *The Great Nutrition Robbery*. New York: Charles Scribner's Sons, 1978.

Hyndman, M.P. (compiler). *Margarine: A Review of Canadian Legislation Relating to Margarine*. Toronto: Institute of Edible Oil Foods, 1961.

Johnston, James P. *A Hundred Years of Eating: Food, Drink and the Daily Diet in Britain Since the Late Nineteenth Century*. Montreal: McGill-Queen's Press, 1977.

Jones, Robert Leslie. *History of Agriculture in Ontario: 1613-1880*. Toronto: University of Toronto Press, 1946.

Kealey, Gregory S. *Hogtown: Working Class in Toronto at the Turn of the Century*. Toronto: New Hogtown Press, 1974.

Lampard, Eric E. *The Rise of the Dairy Industry in Wisconsin: A Study in Agricultural Change, 1820-1920*. Madison, Wis.: State Historical Society, 1963.

Lappé, Frances Moore and Joseph Collins. *World Hunger: Ten Myths*. 4th ed. San Francisco, Calif.: Institute for Food and Development Policy, 1980.

Laskin, Bora. *Laskin's Canadian Constitutional Law; Cases, Text and Notes on Distribution of Legislative Power*. 4th ed., rev. Toronto: Carswell, 1975.

Levere, Trevor H. and Richard Jarrell (eds.). *A Curious Field-book: Science and Society in Canadian History*. Toronto: Oxford University Press, 1974.

Levy, R.L. (ed.). *Nutrition, Lipids and Coronary Heart Disease: A Global View*. Vol. 1: *Nutrition in Health and Disease*. New York: Raven Press, 1979.

Love, E.T. *Oleomargarine and Its Relation to Canadian Economics*. Edmonton: Alberta Dairymen's Association, 1923.

Lloyd's Canadian Food and Packaging Directory. West Hill, Ont.: Lloyd Publications of Canada, [1979].

MacKenzie, David C. *Inside the Atlantic Triangle: Canada and the Entrance of Newfoundland into Confederation 1939-1949*. Toronto: University of Toronto Press, 1986.

Marr, William L. and Donald G. Paterson. *Canada: An Economic History*. Toronto: Macmillan, 1980.

McCallum, John. *Unequal Beginnings: Agriculture and Economic Development in Quebec and Ontario until 1870*. Toronto: University of Toronto Press, 1980.

McIntosh, Dave. *The Collectors: A History of Canadian Customs and Excise*. Toronto: NC Press, 1985.

Mellor, John. *Forgotten Heroes: The Canadians at Dieppe*. Toronto: Methuen, 1975.

Mitchell, Don. *Politics of Food*. Toronto: James Lorimer, 1975.

Mott, Henry Youmans. *Newfoundland Men: A Collection of Biographical Sketches, with Portraits of Sons and Residents*. Concord, N.H.: R.W. and J.F. Cragg, 1894.

Mowat, Farley. *The Regiment*. Toronto: McClelland and Stewart, 1974.

Neatby, H.B. *William Lyon Mackenzie King*. Vol. 3: *1932-39: Prism of Unity*. Toronto: University of Toronto Press, 1976.

Nelson, Jack A. *Hunger for Justice: The Politics of Food and Faith*. Maryknoll, N.Y.: Orbis, 1982.

Newfoundland in the Nineteenth and Twentieth Centuries. Toronto: University of Toronto Press, 1980.

Pabst, Jr., W.R. *Butter and Oleomargarine*. New York: AMS Press, 1980.

Perlin, A.B. *This is Newfoundland*. n.p., n.d.

Pickersgill, J.W. *The Mackenzie King Record*. Toronto: University of Toronto Press, 1970.

Piva, M.J. *The Conditions of the Working Class in Toronto, 1900-1921*. Ottawa: University of Ottawa Press, 1979.

Prentice, E. Parmalee. *Hunger and History: The Influence of Hunger on Human History*. New York: Harper, 1939.

Rea, Kenneth John. *The Prosperous Years: The Economic History of Ontario, 1939-1975*. Toronto: University of Toronto Press, 1986.
Riepma, S.F. *The Story of Margarine*. Washington, D.C.: Public Affairs Press, 1970.
Rowe, Frederick W. *The Smallwood Era*. Toronto: McGraw-Hill, Ryerson, 1985.
———. *A History of Newfoundland and Labrador*. Toronto: McGraw-Hill, Ryerson, 1980.
Ruddick, J.A. *Dairying Industry in Canada*. Ottawa: Department of Agriculture, 1911.
——— et al. *Dairy Industry in Canada*. Toronto: Ryerson, 1937.
Schull, Joseph. *Ontario Since 1867*. Toronto, 1978.
Schüttauf, Werner. *Die Margarine in Deutschland und in der Welt*. Hamburg: Margarine-Union A.G., 1958.
Sheldon, John P. *To Canada, and Through It, with the British Association*. Rev. ed. Ottawa: Department of Agriculture, 1886.
Smallwood, Joseph (ed.). *The Book of Newfoundland*. St. John's: Nfld. Book Publishers, 1967.
Snodgrass, Katherine. *Margarine as a Butter Substitute*. Stanford, Calif.: Stanford University Press, 1930.
Stephenson, H.E. and Carlton McNaught. *The Story of Advertising in Canada*. Toronto: Ryerson, 1940.
Stone, L.O. *Urban Development in Canada*. Ottawa: Queen's Printer, 1961.
Stuijvenberg, Johannes Hermanus van (ed.). *Margarine: An Economic, Social and Scientific History, 1869-1969*. Toronto: University of Toronto Press, 1969.
Tannahill, Reay. *Food in History*. London: Eyre Methuen, 1973.
Thompson, John H. with Allan Seager. *Canada 1922-1939: Decades of Discord*. Toronto: McClelland and Stewart, 1985.
Traill, Catherine Parr. *The Canadian Settler's Guide*. Toronto: McClelland and Stewart, 1969.
Visser, Margaret. *Much Depends on Dinner: The History and Mythology of a Meal*. Toronto: McClelland and Stewart, 1986.
Warnock, John W. *Profit Hungry: The Food Industry in Canada*. Vancouver: New Star, 1978.
Weijs, J.H. *Study of Consumer Purchases and Opinions of Butter and Margarine . . .* (mimeograph). Toronto: Ontario Department of Agriculture, 1961.
Wiebe, Robert. *Search for Order, 1877-1920*. New York: Hill and Wang, 1967.
Wiest, E. *Butter Industry in the United States*. New York: AMS Press, 1980.
Wilson, Charles. *History of Unilever: A Study in Economic Growth and Social Change*. New York: Praeger, 1968.

Articles

Abrahamson, Una. "Margarine (Marine Oil)." *Chatelaine* 43 (February 1970): 14.
Allen, P. "Le Débat Beurre-Margarine au Canada." *L'Actualité Économique* 30 (1954): 420-76.
———. "La Margarine." *L'Actualité Économique* 30 (1954): 187-209.
Ankli, Robert E. and Wendy Millar. "Ontario Agriculture in Transition: The Switch from Wheat to Cheese." *Journal of Economic History* (March 1982): 207-15.
Ball, R.A. and J.R. Lilly. "The Menace of Margarine: The Rise and Fall of a Social Problem." *Social Problems* 29 (1982): 488-98.

Brechin, Maryon. "Consumer Protection." *Encyclopedia Canadiana* Toronto, 1970, 94-98.
"Charts of the Week." *Business Week*, 13 November 1954, 102.
"The Construction and Use of Dietary Standards: A Statement Adopted by the Canadian Council on Nutrition, June 1945." *Canadian Journal of Public Health* (July 1945): 272-75.
Dawson, Helen Jones. "The Consumers' Association of Canada." *Canadian Public Administration* 6 (1963): 92-118.
DeSmith, S.A. "Prohibition of the Manufacture or Sale of Margarine." *Canadian Bar Review* 29 (1951): 206-10.
DeWitt, E. "Oleomargarine Today." *Canadian Welfare* 25 (15 January 1950): 32-34.
"The First Place Margin Narrows." *Business Week*, 22 October 1955, 108 and 110.
Fowke, V.C. "Canadian Agriculture in the Post-War World." *American Academy of Political and Social Science, Annals* 253 (1947): 44-51.
Glass, B (letter). "Oleomargarine Territory." *Science*, 30 September 1966, 1595-96.
Goldstein, Jonah. "Public Interest Groups and Public Policy: The Case of the Consumers' Association of Canada." *Canadian Journal of Political Science* 12 (1979): 137-55.
Hansen, H.A. et al. "Flavour Stability of Canned Butter Concentrate." *Canadian Dairy and Ice Cream Journal* (September 1957): 53-54, 74.
"Heart Motif Puts Margarine One Up On Rival." *Business Week*, 28 January 1961, 98-99.
Howes, H.C. "Can Children Grow up on Oleomargarine?" *Saturday Night*, 20 March 1948, 27.
Kaffezakis, John G. "Margarine: Canada." *Foreign Trade*, 8 August 1964, 11-12.
Lawr, D.A. "The Development of Ontario Farming, 1870-1914: Patterns of Growth and Change." *Ontario History* 64 (1972): 239-51.
Lemoine, J.B. "L'Oleomargarine." *Rélations*, March 1949, 231-33.
Letters. "The Rapeseed Pioneers." *Saturday Night*, May 1980, 11.
Lithwick, N.H. "An Economic Interpretation of the Urban Crisis." *Journal of Canadian Studies* (August 1972): 36-49.
Lloyd, Alan G. "Competition Between Table Margarine and Butter." *Review of Marketing and Agricultural Economics* (March 1936): 5-17.
――――. "Table Margarine in Australia." *Review of Marketing and Agricultural Economics* (March 1951): 24-46.
MacPherson, W.P. "What Has Happened To That 'Butter Shortage'?" *Saturday Night*, 24 May 1941, 30-31.
Manning, T.W. "Agricultural Potentials of Canada's Resources and Technology." *Canadian Journal of Agricultural Economics* (February 1975): 17-29.
"Margarine." *Today's Health* 29 (October 1941): 2.
"Margarine." *Encyclopedia Canadiana*, 6, 381.
"Margarine's Day in Court: Canadian Court Ruling." *Newsweek*, 27 December 1948, 32-33.
"Margarine Offers Important Lesson in Product Development." *Food Engineering*, March 1976, EF9-EF10.
"Margarine Stands on Its Merits." *Science Digest*, April 1950, 68-69.
McGill, A. "The Control of Dyes in Our Food Products." *Industrial Canada*, November 1923, 82.
McInnis, Marvin. "The Changing Structure of Canadian Agriculture 1867-1897." *Journal of Economic History* 42, 1 (March 1982): 191-98.

Melnick, Daniel. "Development of Organoleptically and Nutritionally Improved Margarine Products." *Journal of Home Economics* 60, 10 (December 1968): 793-98.

Miksta, S. "Margarine: 100 Years of Technological & Legal Progress." *Journal of the American Oil & Chemical Society* (April 1971): 169a-72a.

Mouser, E.R. "Diet Margarines: Fat Content, Serving Portions, and Acceptance." *Journal of American Dietetic Association* (January 1969): 29-31.

Nicholls, William H. "Some Economic Aspects of the Margarine Industry." *Journal of Political Economy* 54, 2 (June 1946): 221-42.

"Oleomargarine." *Agricultural and Industrial Progress in Canada* (August 1920): 136, in Canadian Pacific Corporate Archives.

"Oleomargarines." *Annals of Allergy*, April 1970, 164-65.

"Oleo Ads Churn Up New Fight." *Business Week*, 11 June 1955, 70.

Pett, L.P. "Are There Nutritional Problems in Canada?" *Canadian Medical Association Journal* 58, 11 (October 1948): 326-28.

———. "A New Dietary Standard for Canada, 1949." *Canadian Medical Association Journal* 59, 12 (November 1949): 451, 455.

——— et al. "The Development of Dietary Standards." *Canadian Journal of Public Health* (June 1945): 232-39.

"Reference re Validity of Section 5(a) of the Dairy Industry Act." *Dominion Law Reports*. All Canada series, 4 (1950): 689-702.

"The Role of Dietary Fat in Human Health." National Academy of Sciences (U.S.) National Research Council *Publication No. 575*, 1958.

Ruddick, J.A. "The Story of Dairying in Canada." *Canadian Geographical Journal* (July 1937): 41.

Sabry, Zackary et al. "Nutrition Canada." *Nutrition Today* (January-February 1974).

Sahasrabudhe, Medhu R. and C.V. Kurian. "Fatty Acid Composition of Margarines in Canada." *Journal of Canadian Institute of Food Science Technology* 12, 3 (July 1979): 140-43.

"Sale of Margarine a Provincial Matter." *World Affairs* (November 1950): 16.

Smith, M. "What You Should Know About Margarine." *Health* (January-February 1950): 25.

Tisdale, F.F. "Final Report on the Canadian Red Cross Food Parcels for Prisoners-of-War." *Canadian Medical Association Journal* 59, 4 (March 1949): 270-86.

———. "Further Report on the Canadian Red Cross Food Parcels for Prisoners-of-War." *Canadian Medical Association Journal* 55, 3, (February 1944): 135-38.

Weinberg, Bernd. "Trends in Canadian Margarine Legislation." *Journal of the Canadian Institute of Science Food Technology* 7, 1 (January 1974): A12-A14.

Wilson, R. "Laying It On Thick." *Marketing*, 5 March 1984, 2.

Young, James Harvey. "This Greasy Counterfeit: Butter versus Oleo-margarine in the United States Congress, 1886." *Bulletin of the History of Medicine* 53 (1979): 392-414.

Personal Interviews

J.H. Anderson, Saskatoon, Saskatchewan.
Fran Bowden, Charlottetown, P.E.I.
Douglas Busby, Orangeville, Ontario.
S.J. Harris, Ottawa, Ontario.

William Hayward, Waterloo, Ontario.
Hans Heick, Ottawa, Ontario.
J. Mateo, Ottawa, Ontario.
Allan J. Phillips, Toronto, Ontario.
Arthur Schlote, Kitchener, Ontario.
Carl Schlote, Kitchener, Ontario.
William Schlote, Waterloo, Ontario.
David Scott-Atkinson, Toronto, Ontario.
Bryan Stortz, Kitchener, Ontario.
Gerry Stortz, Guelph, Ontario.
James Trouble, Toronto, Ontario.

Miscellaneous

Canadian Annual Digest, 54th Issue, 1949. *Constitutional Law II.*
Canadian Annual Review (various years).
Canadian Broadcasting Corporation. "Dieppe," on *CBC*-TV, 12 November 1979.
———. "Noon Farm Broadcast," 23 July 1982.
Canadian Consumer Survey, 1959-65. Toronto: Canadian Daily Newspaper Publishers' Association, 1963 and 1965.
Canadian Institute of Public Opinion. *The Gallup Report*, 1941-1988.
Canadian Legion members / W.H. Heick correspondence.
Canadian Parliamentary Companion, 1886. Ottawa: J. Durie and Son, 1886).
Chatelaine Consumer Council. *Chatelaine Food Survey: Part 1, Baking, etc.*, 1972.
Cluett, Roslyn. "The Role of Newspapers in the National Margarine Debate, 1946-50." Paper written for History 680 at Wilfrid Laurier University, 1981-1982.
Crosbie, J.C. "A Study of the Vitamin Content of a Butter Substitute Manufactured in Newfoundland." Ontario Agricultural College thesis, 1931.
Gaskell, John. "The Supreme Court of Canada and Section 5(a) of the Dairy Industry Act." Master's cognate essay, Wilfrid Laurier University, 1982-1983.
Haslett, Earl Allan. "Factors in the Growth and Decline of the Cheese Industry in Ontario, 1864-1924," Ph.D. dissertation, University of Toronto, 1969.
Historical Statistics of Canada. Toronto: Macmillan, 1965 and Ottawa: Statistics Canada, 1983.
Ilori, C.O. "Analysis of Retail Store Merchandising Practices of, and Consumer Preferences for, Butter and Margarine," M.Sc. Thesis, University of Toronto, 1958.
Institute of Edible Oil Foods. *Proposal to the Oleomargarine and Butter-Blends Committee, Ontario Ministry of Agriculture and Food, for the Removal of Colour Restrictions on Margarine.* Toronto: The Institute, December 1978.
———. "The Canadian Edible Oil Foods Industries" (Brief submitted to the Royal Commission on Canada's Economic Prospects). Ottawa, 1956.
Joy, John. "The Growth and Development of Trades and Manufacturing in St. John's, 1870-1914." M.A. thesis, Memorial University of Newfoundland, 1977.
Richardson, Carol West. "Responses to Consumerism in Canada: Case Studies of Governmental, Voluntary and Business Responses." M.A. thesis, Carleton University, 1976.
Scott's Directories. Oakville, Ontario, 1985-1986 and 1986-1987.

"Spreading War," on *CHCH*-TV, Hamilton, Ontario, 1985.
Statistics Canada. Cansim University Base, Ottawa, 1986.
Stortz, Bryan. "A History of the Struggle Between Butter and Margarine for the Canadian Market." Essay written in History 680 at Wilfrid Laurier University, 1980-1981.
Stuart, Craig Robert. "Genesis of Federal Food and Drug Legislation in Canada, 1874-1890." Master's cognate essay, Wilfrid Laurier University, 1985.
Thompson, Andrew. "T.L. Kennedy." B.A. thesis, Wilfrid Laurier University, 1980.

INDEX

acids, saturated and unsaturated fatty.
   See also Fats and Oils
advertising, 40
   butter, 154, 156, 157, 158
   margarine, 40, 42, 149, 150, 151, 154, 158, 159
Advertising Standards Council of Canada, 157
Agricultural Price Support Act, 119
Agriculture Canada, 10, 24, 36, 37, 83, 124, 133, 157
Agriculture Prices Support Act, 1944, 95
Agriculture Stabilization Act, 119
   Agricultural Stabilization Board, 121
Agriculturist and Canadian Journal, 6
Alberta, 44, 54, 71, 72, 132
   dairy industry, 24, 25, 67, 94, 100, 101, 112, 113
   margarine regulation, 101, 102, 104, 112, 113, 155
Alberta Dairy Commissioner, 41
Alberta Dairymen's Association, 32, 44, 101, 102
Alberta Federation of Agriculture, 102
Allen, Benjamin, 14
allergies, 141, 142, 144
Altmann, Alfred, 66
American Federation of Labour, in Canada, 45
Anti-Oleo, 44
Archibald, Cyril, 11, 13
Armour and Co., 38, 40, 45, 46
Asseltine, W.M., 64
Associated Producers, 147
Attorney-General of Canada, 81, 82, 83, 84, 85
Ault Foods Ltd., 130
Australia, 10, 29, 54, 56, 63, 68, 78
Aylmer Company, 70

Baldwin, Wilks, 48
Bank of Canada, 119
Barrow, Edgar, 138
Beaulieu, L.E., 82
Bench, J.J., 65, 66, 73, 79
Bengough, Percy, 67
Bennett, Richard Bedford, 55
Bergin, Darbey, 15, 16
Best Foods, 148

Beversdorf, Wally, 128
Bird, T.W., 51
Blake, Edward, 17
Blakeney, Alan, 111
Borden, Robert, 29, 30, 36, 47, 49
Borden and Company, 71
Bourassa, Robert, 107
Bowell, Mackenzie, 13, 14, 17, 19
Bowman, Frank, 104
Bracken, John, 71
Brantford Expositor, 148
Brantford Ontario, City Council, 57, 71
Britain, 2, 7, 8, 9, 10, 11, 15, 24, 25, 30, 31, 41, 43, 51, 55, 56, 63, 68, 119, 122, 125, 133, 138
   and margarine, 29, 35, 36, 38, 58, 65, 67, 76
British American Cultivator, 8
British Columbia, 49, 71, 72, 101, 112, 132, 144, 145, 147, 149
   dairy industry, 94, 100, 102
   margarine regulation, 96, 97, 102, 103 104, 113
British Columbia Federation of Agriculture, 57, 103
British North America Act, 1867, 10, 75, 79, 80, 81, 82, 84, 85, 86, 87, 90, 96
Burns Limited, 147
butter, 1, 2, 7, 8, 9, 12, 16, 25, 27, 29, 30, 31, 33, 34, 38, 43, 84, 85, 95, 105, 106, 119, 134, 162, 163
   blending, 99, 108, 109, 110, 112, 130. *See also* fats and oils
   butter industry, 2, 13, 22, 24, 26, 30, 36, 43, 44, 45, 50, 55, 63, 64, 94, 98, 114, 117-33, 147-59
   colour. *See* colour
   competition with margarine, 2, 3, 14, 21, 22, 29, 31, 32, 40, 43, 48, 63, 64, 69, 75, 95, 99, 106, 130, 131, 143, 147-59
   creamery butter, 28, 36, 41, 50, 53, 70
   dairy butter, 28, 36, 41, 50, 53
   economic comparison with margarine, 115, 128, 129
   exports of, 9, 10, 18, 22, 25, 29, 31, 41, 45, 55

INDEX

health factor, 16, 26, 27, 35, 133-42
legislation re butter, 1874, 11; 1878, 11, 13; 1896, 28, 29; 1903, 28; 1914, 54, 55
lobby, 3, 22, 27, 30, 32; World War I and after, 37, 39, 41, 44, 46, 47, 57; activities re end the ban fight, 1946-50, 67, 71, 72, 80, 94, 103; activities since 1950, 113, 130
nutritional standards of, 26, 27, 133-42
producers of, 3, 10, 27, 37, 41, 57, 64, 77, 90, 107, 131
production of, 3, 9, 23, 57, 63, 118
substitutes, 3, 10, 11, 13, 15, 20, 27, 28, 31, 35, 38, 41, 45, 54, 55, 67, 77, 78, 98, 99, 103, 104, 106, 107, 108, 109, 113, 115, 129, 130, 138, 163; butterine, 12, 20; process or renovated butter, 28, 50
technology, 8, 9, 129, 130

Caldwell, George, 155
Calgary Herald, 38
Camp, Dalton, 113
Campbell, Douglas, 96
Canada, agriculture, 3
commissioner of Dairy and Cold Storage, 41
federal legislation, 105; 1878, 11; 1886, 18-22, 82, 85, 87, 88, 97; 1917, 37; 1927, 83, 85; 1951, 95, 96, 97
Canada Food Board, 31, 36
Canada Food Guide, 137, 143
Canada Packers, 66, 101, 130, 147, 148, 151
Canadian Annual Review, 43, 50
Canadian Bank of Commerce, 63
Canadian Brotherhood of Railway Workers, Ladies' Auxiliary, 109
Canadian Chamber of Commerce, 67
Canadian Congress of Labour, 67
Canadian Council of Agriculture, 10, 50, 64
Canadian Council on Nutrition, 136
Canadian Dairy and Ice Cream Journal, 69

Canadian Dairy Commission, 121, 122, 157
Canadian Federation of Agriculture, 42, 68, 72, 80, 82, 89, 90, 117, 118, 153
Canadian Grocer, 46
Canadian Home Journal, 149
Canadian Legion, 102, 103
Canadian Manufacturer's Association, 80
Canadian Medical Association, 33, 35, 67, 155
Journal, 33, 35, 67, 83
Canadian National Exhibition, 42
Canadian Pacific Railroad, 46
Canadian Red Cross, 58
canola, 128, 142, 162. *See also* rapeseed and fats and oils
Carling, John, 14
Casey, George, 17
Cattle Breeders' Association, 32
Central Alberta Dairy Pool, 66
Chatelaine, 156
cheese, 7, 8, 16, 23, 25, 29, 31, 43, 54, 56, 63, 68, 84, 95, 112, 113, 118, 121, 122, 158
Chemical Institute of Canada, 71
Chevreul, Michel, 1
Claxton, Brooke, 75
colour, 2, 9, 16, 42, 73, 76, 91, 97, 115, 122, 129, 142, 143-46, 148, 153, 156, 157
colour regulations during World War I, 32, 35, 37, 48
provincial colour regulations: Alberta, 101, 102, 108, 112, 155; British Columbia, 103, 104; Manitoba, 108, 110; New Brunswick, 99, 108, 109, 110; Nova Scotia, 100, 108; Ontario, 98, 99, 106, 108, 113, 114, 115, 143, 144, 151; Quebec, 107, 108, 143; Prince Edward Island, 107, 108; Saskatchewan, 101, 155
Colborne, F.C., 102
Coldwell, M.J., 76
Conroy, Pat, 67
Conseil de l'industrie Latière du Québec, 108
Consumer and Corporate Affairs, Department of, 157

Consumers' Association of Canada, 67, 80, 82, 85, 105, 109, 145, 146
Consumers' Association of Ontario, 130
Consumer Reports, 154
consumers of butter and margarine, 2, 4, 11, 14, 15, 16, 17, 25, 27, 32, 33, 39, 40, 43, 45, 46, 47, 48, 55, 56, 63, 64, 70, 73, 79, 88, 97, 99, 100, 101, 107, 108, 109, 110, 111, 112, 122, 128, 130, 145, 147, 149, 151, 159, 163
Cooperative Commonwealth Federation, 72, 76, 101. *See also* New Democratic Party
Costigan, John, 13, 14, 19, 20
Coughlin, Bing, 58
cows, 1, 8, 22, 23, 43, 53, 56, 63, 70, 77, 83, 112, 117, 119, 131, 152, 162
Crerar, Thomas, 31, 32, 37, 42, 48
Croll, David, 96
Cruikshank, George, 76, 95, 96
Cummings, Jane, 155
Curtis, C.A., 71

Dairy Bureau of Canada, 142, 144, 145, 156, 157, 158
Dairy Farmers of Canada, 67, 72, 77, 90, 154
dairy industry, 3, 8, 10, 12, 14, 16, 17, 23, 24, 25, 26, 27, 28, 29, 36, 37, 39, 41, 42, 43, 44, 45, 47, 48, 50, 53, 55, 62, 63, 64, 65, 66, 68, 73, 81, 84, 87, 94, 96, 99, 110, 112, 118, 130, 144, 147, 152, 153, 162
  creameries, 23, 24, 50
  legislation, dairy: 1874, 11; 1878, 11; 1896, 28, 29; 1903, 27; 1914, 27, 29; 1927, 73, 80, 83, 84, 87; 1951, 95, 96
Dairymen's Association of Western Ontario, 12
Davies, William, Company, 38
Davis, William, 113, 115, 146
Denmark, 2, 10, 24, 32, 45, 78, 133, 163
Depression, 53, 54, 55, 56, 62, 68, 70, 119
Digby Weekly Courier, 37

Dominion Dairies, 130
Douglas, T.C., 101, 111, 112
Downey, R.K., 128
Drew, George, 71, 99
Dunning, Charles, 54
Duplan, J.H., 89, 152
Duplessis, Maurice, 98, 107

Estey, Willard, 83, 86, 87
Euler, William Daum, 3, 29, 47, 49, 50, 63, 64, 150
  Euler and others, 80, 82, 83
  fight to end the ban, 62-78, 79, 84, 88, 89, 90, 95, 105, 138, 163
Europe, 1, 2, 5, 10, 35, 40, 43, 44, 109, 117, 132, 164
European Recovery Plan (Marshall Plan), 70

Farm Radio Forum, 72
Farmer's Advocate, 11, 12, 17, 27, 29, 31, 32, 35, 44, 63, 71, 96
Farmilo, Alfred, 45, 46
fats and oils, 1, 2, 10, 15, 27, 36, 39, 46, 55, 57, 65, 69, 73, 107, 108, 110, 111, 113, 114, 117, 118, 125, 128, 129, 132, 138, 141, 143, 147, 156, 162
  cholesterol, 141
  hydrogenation process, 27, 39, 148, 157
  seal oil, 125
  vegetable oils: coconut, 132; corn, 132, 159; cottonseed, 132; palm, 132; soybean, 114, 119, 125, 128, 129, 132, 162; sunflower, 128, 148, 159. *See also* canola and rapeseed
Fielding, W.S., 13, 49
Fisher, Sidney, 14, 17, 28
Fisheries Council of Canada, 105
Flemming, Hugh John, 109
food, 1, 36, 70, 77, 105, 117, 118, 162, 163
  adulteration, 11
  vitamins. *See* nutrition
  and World War I, 29-42
food and drug regulations, 11, 101
Food Prices Review Board, 137

INDEX

France, 1, 9
Free Press (Ottawa), 18
Frost, Leslie, 99, 114, 146

Gallup Poll, 67, 78, 98
Gardiner, James, 66, 75, 95, 96
Gazette (Montreal), 65
General Agreement on Tariffs and Trade, 66, 74, 75, 96
Germany, 10, 58, 144
Gillmor, Arthur, 14
Globe (Toronto), 19
Globe and Mail (Toronto), 68, 75
Good, William C., 49
Great War Veterans' Association, 46
Green Giant, 156
Gregg, Milton, 75
Grocery Products Manufacturers of Canada, 108
Guardian (Charlottetown), 37, 97, 107
Guillet, George, 16

Hahn, Joyce, 89
Haig, John, 64
Hanley, Frank, 98
Hanna, W.J., 31, 32, 37
Hannam, H.H., 68, 69, 72, 89, 153
Hardy, A.C., 57, 72, 73
Harris Abbatoir, 38
Harvestore Systems, 157
Hayden, Salter, 82, 90
Health and Welfare Canada, 76, 83, 89, 104, 105, 137, 143, 156
Henderson, Lorne, 130
Herald (Halifax), 31, 73
 Chronicle-Herald, 100
Hesson, Samuel, 16
Hickey, Charles, 17
Holland. *See* the Netherlands
Homemaker's Clubs, 44
Homer, Brian, 156
Hospital Council of Canada, 71
Hougen, Frithjol, 128
Hydrogenation process. *See* fats and oils
Hyndman, Margaret, 82, 83

Imperial Order of the Daughters of the Empire, 46

Industrial Revolution, 1
Industry, Trade and Commerce, 95
Institute of Edible Oil Foods, 105, 106, 114, 145, 146, 150, 152, 155, 157
Ireland, 9
Irvine, William, 46
Italy, 58

Jackson, Joseph, 15
Jelke, 148
Jenkins, John, 14
Journal (Stanstead), 18
Judicial Committee of the Privy Council, 79, 80, 85, 86, 88, 89, 90, 91, 97, 98, 147, 152, 153

Keefer, Francis H., 42
Kellock, Roy, 83, 86
Kennedy, T.L., 98
Kerwin, Patrick, 80, 84, 85
Keys, Ansel, 155
Killick, C.H.P., 72
King, W.L. Mackenzie, 45, 47, 49, 50, 51, 54, 55, 63
 involvement in campaign to end the ban, 64, 65, 66, 74, 75, 76, 78, 79, 83, 88, 89, 95
Kitchen, Earle, 72
Kitchener Daily Record, 65, 159
Knowles, Stanley, 106
Kraft Foods, Ltd., 66, 148, 150, 151, 158

Langevin, Hector, 20
L'Association Canadienne des Electrices, 68, 80, 84, 85
Labour Council of Montreal, 35
Lambert, N.P., 64
Laurier, Sir Wilfrid, 28, 30
Leisemer, A.J., 112
Lever Brothers, 90, 99, 151
Liberal Party of Canada, 45, 47, 49, 51, 52, 54, 66, 71, 73, 76
 of Ontario, 115, 146
Locke, Charles, 83, 85, 86
Love, E.T., 44, 45
Lumbermen's Association, 71
L'Union des Cantons de L'Est, 18

Macdonald, Sir John A., 13
Machado, Sonza, 128
Mackenzie, Ian, 73
Maclean's, 83, 96, 149
MacLeod Gazette, 19
Manitoba, 28, 54, 71, 72, 96, 132
  dairy industry, 94, 100
  margarine regulation, 100, 104, 110
Manning, Ernest, 101
margarine
  and armed forces: World War I, 38; World War II, 57-60, 75
  ban on, 2, 3, 12, 14, 17, 19, 20, 21, 22, 28, 29, 30, 35, 36, 37, 38, 42, 43, 48, 49, 51, 52, 53, 55, 60, 63, 65, 66, 68, 73, 74, 72-91, 95, 97, 98, 101, 103, 109, 113, 114, 133, 146; ban by Prince Edward Island, 97, 106; ban by Quebec, 97, 104, 107
  blends. *See* butter, blending
  brands of, 38, 39, 40, 107, 148, 149, 150, 151, 152, 155, 158, 159
  colour. *See* colour
  competition with butter, 2, 14, 21, 22, 36, 40, 43, 46, 55, 64, 69, 75, 95, 99, 106, 130, 131, 143, 147-59
  diet margarine, 107, 110
  economic comparison with butter, 115, 128, 129
  federal legislation, 3; 1878, 11; 1886, 18-21, 82, 85, 87, 88, 97, 143, 147; 1917, 37; 1927, 83, 85; 1951, 95, 96, 97
  health and, 12, 20, 53, 82, 83, 85, 110, 115, 133-42, 148, 149, 150, 152, 153, 154, 155, 156, 158
  importation, 3, 13, 14, 18, 19, 20, 37, 50, 73, 84, 85, 86, 90, 95, 96, 104, 105, 119, 120, 133; and export, 128
  industry, 12, 16, 17, 18, 27, 29, 32, 37, 39, 40, 42, 43, 44, 45, 49, 90, 95, 97, 98, 117-33, 138, 145, 147, 152, 153
  legalization fight, 1946-50, 71-78, 79-91, 96, 97, 138, 143
  legalized, 1917-24, 30, 31, 35, 36, 37, 38, 48-53, 54, 64, 74, 82, 149
  lobby, 46, 47, 64, 65, 80, 105
  nutritional standards of, 39, 44, 55, 115, 134-43
  margarine reference, 79-91
  origins, 1
  taste, 41, 42, 148, 149
  technology of, 2, 12, 27, 39, 125, 126, 127, 128, 143
Maritime Farmer, 26, 37
Martin, Paul, 89
Massey, C.A., 151
McDonalds, 156
McLelan, Archibald, 12, 13, 14, 15, 19
McMillan, Gilbert, 72
McMullen, James, 14
Mége-Mouriez, Hippolyte, 1
Meighen, Arthur, 47, 48, 49, 50, 54, 76
Mercier, Honoré, 13
Messer, J.R., 111, 112
milk, 8, 9, 22, 23, 24, 25, 26, 27, 28, 30, 31, 36, 43, 48, 50, 54, 56, 63, 64, 65, 68, 70, 75, 77, 81, 84, 100, 101, 108, 109, 117, 118, 119, 122, 134, 138, 154, 156, 162. *See* myth
Milliken, R.H., 82
Mills, David, 17
Monarch Fine Foods, Ltd., 157
Moncton Daily Times, 37
Monetary Times, 18
Montreal Star, 144
Motherwell, William R., 49, 50, 54
Mowat, Farley, 58
Munro, Ross, 58
Murdoch, James, 73
Murray, Keith, 157
Mustard, Fraser, 156
myth, agrarian, 2, 3, 21, 26, 27, 42, 43, 65, 109, 134, 146, 156, 159
  of colour, 143, 144
  of milk, 2, 16, 18, 26, 33, 42, 44, 46, 47, 53, 56, 109, 134, 143, 158, 163

Nabisco, 157
Napoleon III, 1
National Council of Women, 44, 46, 71, 74, 103, 109
National Dairy Council of Canada, 54, 77, 78, 80, 89, 94, 96, 130, 152

National Dairymen's Association (Canada), 50
National Dairyman's Association (U.S.A.), 66
National Research Council, 128
Nealy, D.B., 35
Neill, A.W., 49, 51
Nesbitt, E.W., 29, 42
Netherlands, the, 2, 10, 60, 133, 138
New Brunswick, 6, 7, 14, 18, 22, 71, 72, 151
  dairy industry, 94, 99, 109
  margarine regulation, 96, 99, 100, 104, 109, 110
New Brunswick Cream Producer's Marketing Board, 109
New Brunswick Dairymen's Association, 109
New Democratic Party, Ontario, 146
Newfoundland, 66, 69, 75, 79, 95, 105, 131
  dairy industry, 94
  lack of margarine regulation, 75, 97, 104, 106
  margarine industry, 10
Newman, William, 145
New Zealand, 24, 29, 54, 56, 63, 68, 74, 76, 78, 119, 133, 163
Nollet, I.C., 111
Norway, 130
North West Territories, 13, 104
Nova Scotia, 6, 7, 12, 13, 14, 18, 22, 27, 44, 71, 72, 132, 151
  dairy industry, 94, 99
  margarine regulation, 96, 99, 100, 104, 108, 109, 130
Nova Scotia Dairy Commission, 109
Nova Scotia Milk and Cream Producers, 109
nutrition, 3, 16, 26, 30, 33, 54, 56, 62, 64, 67, 75, 76, 91, 105, 122, 134-42
  Standing Committee on Nutrition, 58
  vitamins, 26, 27, 34, 35, 45, 55, 136, 137, 138, 147, 148, 151, 157, 162

Observer (Sarnia), 18
oils. *See* fats and oils

oleomargarine, 1, 2, 11, 12, 15, 26, 27, 37, 39, 45, 46, 47, 51, 66, 71, 104, 115. *See also* margarine
Ontario, 3, 6, 7, 14, 15, 16, 18, 21, 24, 28, 42, 48, 71, 72, 101, 129, 130, 132, 145, 148, 151, 153
  dairy industry, 9, 21, 22, 94, 113, 121, 122
  regulation of margarine, 3, 11, 98, 99, 104, 106, 113, 114, 143, 144, 151
Ontario Agricultural and Arts Association, 12
Ontario Agricultural Commission, 9
Ontario Cheese Producers Marketing Board, 77
Ontario Concentrated Milk Producers, 114
Ontario Cream Producers Association, 89
Ontario Creamery Association, 114
Ontario Dairy Council, 130
Ontario Federation of Agriculture, 99, 113
Ontario Federation of Home and School Associations, 71
Ontario Firefighters' Federation, 71
Ontario Heart Foundation, 157
Ontario Margarine Committee, 114, 146
Ontario Milk Marketing Board, 158
Ontario Research Foundation, 129
Ottawa Anti-Tuberculosis Association, 35
Ottawa Board of Control, 46

Pagé, Michel, 115
Parry, Peter, 89
Paterson, William, 17, 18, 19, 73
Pearson, Lester, 121
"Penny Wise," 103, 113
Peterson, David, 115, 146
Power, Lawrence, 29
Prairie Vegetable Oils Ltd., 125
Prince Edward Island, 6, 7, 10, 14, 18, 55, 71, 72, 94, 153
  dairy industry, 107
  margarine regulation 96, 97, 98, 99, 104, 106, 107, 108

228   THE PROPENSITY TO PROTECT

Progressive Conservative Party, 45, 47, 52, 66, 71, 73, 76
   Ontario, 113, 115, 146
Progressive Party, 45, 47, 49, 50, 51, 52
protective impulse, 3, 5, 9, 18, 19, 21, 27, 29, 41, 42, 45, 46, 47, 48, 51, 97, 102, 103, 115, 134, 144, 146, 154
Putnam, Frank, 103

Quebec, 3, 6, 7, 8, 13, 14, 18, 71, 106, 129, 130
   arguments before the Supreme Court and the Judicial Committee of the Privy Council, 80, 82
   dairy industry, 9, 22, 24, 31, 72, 94, 97, 107, 108, 115, 121, 145
   margarine legislation, 3, 97, 98, 99, 104, 107, 108, 115, 143
Quebec Food Council, 108

Rand, Ivan, 83, 87
rapeseed, 111, 112, 119, 125, 128, 142. *See also* canola
Rapeseed Association of Canada, 105
Reardon, C.H., 108, 109
reciprocity, 1911, 28
Reed, Percy, 44, 55
Reilly, Leonard, 114
Retail Merchants Association, 56
Rinfret, Thibodeau, 83, 84, 85, 90
Robertson, Wishart, 65, 72
Roblin, Duff, 110
Roebuck, Arthur, 73
Rolston, Peter, 113
Rolston, Tillie, 103, 104, 113
Romahn, James, 159
Royal Commission on Canada's Economic Prospects, 117
Royal Commission on Taxation (1963-66), 106

Sara Lee, 156
Saskatchewan, 25, 35, 54, 70, 72, 154
   dairy industry, 94, 100, 111
   margarine regulation, 96, 100, 104, 111, 130, 155

Saskatchewan Dairy Association, 31, 55, 66, 101
Saskatchewan Dairy Commissioner, 41, 44
Saskatchewan Dairy Pool, 238
Saskatchewan Department of Agriculture, 54
Saskatchewan Federated Cooperatives, 111, 129, 153
Saskatchewan Grain Growers' Association, 32
Scott-Atkinson, David, 155
Scott-Atkinson, Only International, Ltd., 155
Shortt, Elizabeth, 54
Shuswap-Okanagan Dairy Industries Cooperative Association, 103
Sinclair, James, 71, 72, 74, 76, 77
Sinclair, Upton, *The Jungle*, 14
Snodgrass, Katherine, 2
Social Credit Party, 72, 76, 101, 102
Soya Bean Grower's Marketing Board, 130
Stacey Brothers, 130
Standard Brands, Ltd., 148, 155
Statistics Canada, 83
Stefansson, Baldur, 128
St. Laurent, Louis, 74, 76, 95, 96
Supreme Court, 73, 74, 75, 77-91, 94, 96, 97, 98, 100, 102, 125, 148, 152, 153
Sutherland, Donald, 48
Sweden, 130
Swift Canadian, 38, 39, 45, 46, 66, 101, 147, 151

Taschereau, Robert, 83, 87
Taylor, A.C., 99
Taylor, George, 11, 13, 14, 15, 16, 17, 20
Taylor, Gordon, 112
Thatcher, Ross, 71, 111
Tolmie, Simon, F., 49
Toronto City Council, 56
Toronto Star, 96
Trades and Labour Council, 45
Traill, Catherine Parr, 8

Trow, James, 18, 19
Turrell, J.G., 35

Unilever, 55, 150, 151. *See also* Lever Brothers
United Farmers of Alberta, 32, 46
United Farmers of British Columbia, 32
United Nations, 105
United States, 2, 7, 8, 9, 14, 15, 16, 24, 28, 31, 50, 64, 70, 77, 117, 125, 132, 144, 150, 158
   dairy industry, 25, 35, 51, 63
   legislation, 12, 18, 95, 106, 147
   margarine, 10, 11, 13, 40, 42, 58, 65, 69, 73, 115, 120, 151, 153
Unwin, George, 153

Ure, David, 102, 112

Vancouver News-Herald, 103
Vancouver Junior Camber of Commerce, 103
Varcoe, Frederick, 81, 82, 85
Venoit, P.J., 64

Waines, William J., report by, 110, 125
war brides, 60, 65
Warner, M.D., 71
Wartime Prices and Trade Board, 56, 57, 70, 122
Welfare Council of Canada, 71
wheat, 3, 6, 7, 8, 9, 14, 22, 31, 118, 119, 128, 137
White Paper on Unemployment and Income, 69
Wilder, F.W., 39
Wilson, W.A., 32, 36, 41, 42, 47
Winch, E.E., 104
Winnipeg Free Press, 19, 50
Winnipeg Tribune, 110
Wismer, G.S., 102, 103
Women's Christian Temperance Union, 46
Women's Institutes of Canada, 44, 100
World's Columbian Exposition, 23
World War I, 22, 25, 26, 28, 29–42, 53, 54, 57, 83, 90, 132, 136, 143, 149

World War II, 3, 47, 53–60, 62, 63, 119, 122, 124, 125, 135, 136, 138, 148, 149, 152
Wren, Albert, 114

Yukon, 104